©2005 Algrove Publishing Limited
ALL RIGHTS RESERVED.
No part of this book may be reproduced in any form, including photocopying, without permission in writing from the publishers, except by a reviewer who may quote brief passages in a magazine or newspaper or on radio or television.

Algrove Publishing Limited
36 Mill Street, P.O. Box 1238
Almonte, Ontario, Canada K0A 1A0

Telephone: (613) 256-0350
Fax: (613) 256-0360
Email: sales@algrove.com

Library and Archives Canada Cataloguing in Publication

     Chain saw and crosscut saw training course : student's guidebook / United States Department of Agriculture, Forest Service, Technology & Development Program. -- 2001 ed.

ISBN 978-1-897030-21-9

     1. Chain saws. 2. Crosscut saws. 3. Tree felling.
4. Loggers--Training of. I. Technology & Development Program (U.S.)

TJ1233.C528 2004   634.9'8'0284   C2005-900158-5

Printed in Canada
#3-8-12

Introduction

As a student guidebook produced by the USDA Forest Services, this material is aimed at forestry workers who would normally be working in groups. Accordingly, the information frequently refers to the safety of co-workers. For those working alone in the bush, some of the material will be redundant but most of it applies.

Like much material written today, the extreme emphasis on safety sometimes leads to impractical statements. The best example in this book is the statement in a section on axes that reads "To prevent blows from glancing, keep the striking angle of the tool head perpendicular to the tree trunk." If you actually did that, you could never cut a notch or remove a limb.

In another area, it states "Never chop cross-handed; always use a natural striking action." Well, sometimes the natural striking action is cross-handed. Sometimes you can get at only one side of a tree so you have no choice. It would be more useful to state that you should be conscious of the reduced control you might have when chopping cross-handed and to protect yourself accordingly.

Unfortunately, this latter statement could leave you open to a lawsuit if you are an employer. In real life, forestry workers know they have to chop cross-handed at times and most develop the skill to the point that it is a natural striking action.

So, as you read this text please remember that there is a sprinkling of employer language in it. But there is also much material here that can make you more effective in the bush at the same time as it helps you keep your body intact.

Leonard G. Lee, Publisher
Almonte, Ontario
January 2005

United States
Department of
Agriculture

Forest Service

Technology & Development Program

6700 Safety & Health
July 2001
0167-2815-MTDC

Chain Saw and Crosscut Saw Training Course

Student's Guidebook
2001 Edition

FINAL DRAFT

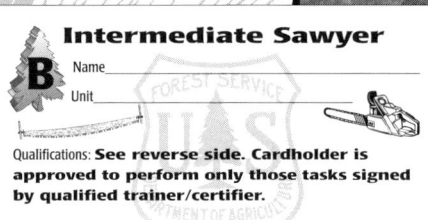

Missoula Technology & Development Center

Algrove Publishing~Classic Reprint Series

Chain Saw and Crosscut Saw Training Course

Student's Guidebook
2001 Edition

FINAL DRAFT

Chuck Whitlock, *Project Leader*

R.C. Carroll, *Northern Region Chain Saw Program Coordinator*

Paul Chamberlin, *Northern Rockies Fireline Safety Specialist*

David Michael, *Pacific Southwest Region Crosscut Saw Coordinator and Trails Specialist*

Winston Rall, *Pacific Northwest Region Chain Saw/Crosscut Program Manager and Occupational Safety and Health Specialist*

Jerry Taylor Wolf, *Project Assistant*

**USDA Forest Service
Technology and Development Program
Missoula, MT**

7E72E41—Chain Saw and Crosscut Saw Training

July 2001

The Forest Service, United States Department of Agriculture (USDA), has developed this information for the guidance of its employees, its contractors, and its cooperating Federal and State agencies, and is not responsible for the interpretation or use of this information by anyone except its own employees. The use of trade, firm, or corporation names in this document is for the information and convenience of the reader, and does not constitute an endorsement by the Department of any product or service to the exclusion of others that may be suitable.

The U.S. Department of Agriculture (USDA) prohibits discrimination in all its programs and activities on the basis of race, color, national origin, sex, religion, age, disability, political beliefs, sexual orientation, or marital or family status. (Not all prohibited bases apply to all programs.) Persons with disabilities who require alternative means for communication of program information (Braille, large print, audiotape, etc.) should contact USDA's TARGET Center at (202) 720–2600 (voice and TDD). To file a complaint of discrimination, write USDA, Director, Office of Civil Rights, Room 326-W, Whitten Building, 1400 Independence Avenue SW, Washington, D.C. 20250–9410, or call (202) 720–5964 (voice and TDD). USDA is an equal opportunity provider and employer.

Contents

Chapter 1—Course Information and Safety Requirements 1
   Forest Service Chain Saw and Crosscut Saw Program 1
   Job Hazard Analysis 1
      First Aid 2
      Emergency Evacuation Plan 2
      Personal Protective Equipment 2
         General Requirements 2
         Specific Requirements 3
   Situational Awareness 5
      Checklist of Personal Safety Considerations and Attitude 5
      Evaluating the Complexity of the Assignment 5

Glossary 6

Sample Job Hazard Analysis 9

Additional Information for Sawyers 13

Chapter 2—Chain Saw Use and Maintenance 14
   Chain Saw Operation 14
      Chain Saw Components 14
      Guide Bar Maintenance 17
      Chain Tension 18
      Daily Saw Maintenance 18
         Cleaning Exercise 19
      Chain Maintenance 19
      Chain Filing 20
         Sharpening Cutters With a Round File 20
         How to Set Depth Gauges 21
         Chain Filing Exercise 22
      Saw Transportation 22
         Transporting Chain Saws in a Vehicle 22
         Transporting Chain Saws by Hand 22
      Safe Chain Saw Use 22
         Starting Procedures 22
         Starting the Chain Saw on the Ground 23
      Operational Safety 23
         Handling 23
         Reactive Forces 23
      Additional Tools 25
         Axes 25
         Wedges 25
         Fuel and Oil Containers 26
         Peaveys and Cant Hooks 27

Chapter 3—Chain Saw Tasks and Techniques 28
   Safe Chain Saw Use 28
      Proper Use of Bumper Spikes (Dogs) 28
   Bucking 28
      Situational Awareness 28
      Safe and Efficient Bucking Techniques 28
         Determining Bind 29
      Safe Bucking Practices 31
         Points to Remember 32

Limbing	32
Brushing and Slashing	33
Sizeup and Safety Considerations	33
Safe and Efficient Brushing and Slashing Techniques	33
Basic Felling	35
Situational Awareness	35
Sizeup	36
Escape Routes	37
Felling the Tree	38
Felling Details	39
The Undercut, Holding Wood, and Backcut	39
Directional Felling	42
Felling Observers and Spotters	43

Chapter 4—Crosscut Saw Tasks and Techniques — 44

Understanding Your Crosscut Saw	44
Historical Origin of the Crosscut Saw	44
Different Types of Crosscut Saws	44
One-Person Crosscut Saws	44
Two-Person Crosscut Saws	44
Two-Person Crosscut Saw Patterns	44
Felling Saws	44
Bucking Saws	44
Saw Grinds	45
Flat Ground	45
Straight Taper Ground	45
Crescent Taper Ground	45
How a Saw Cuts	46
Cutter Teeth	46
Rakers	46
Gullets	47
Tooth Patterns	47
Plain-Tooth (Peg-Tooth) Pattern	47
M-Tooth Pattern	47
Great American-Tooth Pattern	47
Champion-Tooth Pattern	47
Perforated Lance-Tooth Pattern	47
Lance-Tooth Pattern	48
Saw Handles	48
Handle Position	48
Handle Attachment Holes	48
Types of Saw Handles	48
Handle Installation and Maintenance	49
Saw Maintenance	49
Cleaning the Saw	49
Checking for Straightness	50
Testing the Saw	50
Brief Overview of Saw Filing Procedure	50
Storage	51
Transporting Saws	52
Saw-Related Tools and Equipment	52
Lubricants	53
Types	53
Functions	53
Applying Lubricants	53

Contents

Axes	53
Wedges	53
Splitting Wedges	54
Lifting Wedges	54
Peaveys and Cant Hooks	54
Underbucks	54
Types of Underbucks	54
Bucking and Felling Preparation and Techniques	55
Bucking	55
Safety Considerations	55
Bucking Sizeup	55
Planning the Cut	56
Single-Bucking Techniques	59
Single Bucking With No Bind: Top Cutting	60
Single Bucking With Top Bind: Underbucking Required	60
Underbucking	60
Single Bucking With Top Bind: Top Cutting	61
Single Bucking With Bottom Bind: Top Cutting	61
Single Bucking With Bottom Bind: Underbucking	61
Single Bucking With End Bind	61
Single Bucking With Side Bind	61
Double-Bucking Techniques	62
Felling	62
Safety Considerations	62
Felling Sizeup	63
Establishing Escape Routes	64
Placing the Undercut	64
Cutting the Backcut	67

Acknowledgments

Project leader Chuck Whitlock wishes to thank the following individuals for their contribution to this project.

Gary Hoshide, Bert Lindler, Sara Lustgraaf, Emily Ranf, Michelle Beneitone, Bob Beckley, Mark Wiggins, and Ben Croft.

Thanks to everyone who provided input and reviewed the document.

Chapter 1—Course Information and Safety Requirements
(Suggested time: 2 hours)

In this chapter:

• Students will receive an overview of the chain saw and crosscut saw training course.

• Students will be provided with information on the requirements for successfully completing the chain saw and crosscut saw training course.

• Students will be able to identify the elements in a job hazard analysis (JHA): the task or procedure to be accomplished, the hazards associated with the task or procedure, abatement actions to eliminate or reduce the hazards, first-aid supplies, and emergency evacuation procedures.

• Students will learn why personal protective equipment (PPE) is used, how it is used, and how it is maintained.

• Students will be able to identify common safety mistakes made by sawyers.

Forest Service Chain Saw and Crosscut Saw Program

As Forest Service employees you must be aware of all laws and standards that must be met before you operate a chain saw.

Why do we have a national chain saw and crosscut saw program?

This course will provide the skills to safely use chain saws and crosscut saws, and serve as a refresher class for persons who have already completed the training. The national chain saw and crosscut saw program was developed to provide all sawyers a solid foundation for safe and efficient saw operation while felling, bucking, brushing, or limbing.

Safety is the most critical objective. Your safety, the safety of your coworkers, the safety of the public, and property protection should be a part of every plan and every action you take. Careful study and practice of saw operations will improve your own abilities and help you identify your limitations to ensure safe saw operation.

This course is designed to train beginning and intermediate sawyers to perform project work safely and efficiently. The chapters for both courses are summarized below.

The Chain Saw Course:
• Chapter 1 (classroom), course information and safety requirements
• Chapter 2 (classroom), chain saw use and maintenance
• Chapter 3 (classroom), chain saw tasks and techniques
• Chapter 5 (field), sawyer evaluation process and sawyer evaluation form

The instructor will describe the certification levels and details of restrictions or endorsements for special uses.

The Crosscut Saw Course:
• Chapter 1 (classroom), course information and safety requirements
• Chapter 4 (classroom), crosscut saw tasks and techniques
• Chapter 5 (field), sawyer evaluation process and sawyer evaluation form

The instructor will describe the certification levels and details of restrictions or endorsements for special uses.

Job Hazard Analysis

A JHA (see Sample Job Hazard Analysis section at the end of this chapter) must be prepared (preferably with the assistance of the involved employees) before beginning any work project or activity. The JHA must:

• Identify the task or procedure to be accomplished. Such tasks could include limbing, bucking, or felling.

• Identify the hazards associated with the task or procedure. These hazards may include physical, biological, environmental, chemical, and other hazards. Examples of hazards include:
 —Physical hazards: Rocky terrain, slippery slopes.
 —Biological hazards: Insect bites, hantavirus.
 —Environmental hazards: Weather-related hazards such as hypothermia, wind, lightning.
 —Chemical hazards: Hazardous materials such as fuel mix for chain saws and oil for crosscut saws.
 —Other hazards: Personal security issues, public traffic, hunting seasons.

• Identify abatement actions that can eliminate or reduce hazards. Abatement actions include:
 —Engineering controls: The most desirable method of abatement (such as ergonomic tools and equipment).
 —Substitution: Such as switching to high flashpoint, nontoxic solvents.
 —Administrative controls: Such as limiting exposure by reducing work schedules or establishing appropriate work practices and procedures.
 —PPE: The last method of abatement (such as using hearing protection when working with chain saws).

- Identify first-aid supplies and emergency evacuation procedures. In the event of an emergency evacuation, be prepared to provide the following information:
    - Nature of the accident or injury (avoid using the victim's name).
    - Type of assistance needed (ground, air, or water evacuation).
    - Location where the accident occurred, best access to the work site (road name or number).
    - Radio frequencies.
    - Contact person.
    - Local hazards to ground vehicles or aviation.
    - Weather conditions (windspeed and direction, visibility, temperature).
    - Topography.
    - Number of individuals to be transported.
    - Estimated weight of individuals for air or water evacuation.

First Aid

Refer to the *Health and Safety Code Handbook* chapter 20, sections 21.21 and 21.22, for information on handling a medical emergency. The onsite first-aid kit must have supplies that meet Occupational Safety and Health Administration (OSHA) specifications and requirements. A Type IV first-aid kit must be available as a minimum (General Services Administration national stock number NSN 6545-01-010-7754). A more complete kit meeting higher standards may be used.

Emergency Evacuation Plan

An emergency evacuation plan is essential for any field project, especially one involving chain saws and crosscut saws. All employees need to be proficient in using a radio. They need to know which frequencies to use and whom to contact in the event of an emergency. The latitude and longitude and/or the legal location for an emergency medical helispot shall be determined and included in the JHA before starting any work. The entire crew shall know where the helispot is located. The emergency evacuation plan needs to be updated when the work location changes.

The JHA and emergency evacuation plan shall be signed by employees, signifying that they have read and understood the contents, have received the required training, are qualified to perform the task or procedure, and will comply with all safety procedures.

A copy of the JHA, the bloodborne-pathogen exposure control plan, the material safety data sheets for products used on the work project or activity, and the emergency evacuation plan must be kept onsite during the project. The JHA can be reviewed and updated during tailgate safety sessions. These sessions take place before a new project or activity is begun, when changes are made (such as changing location, adding crewmembers, or changing job responsibilities), or whenever employees believe a session is needed. Topics often focus on the hazards associated with the job and methods to eliminate or abate them.

Personal Protective Equipment
(*Health and Safety Code Handbook* chapter 70, section 72)

Items that must be included in the JHA:

(Chain Saw Operations)	(Crosscut Saw Operations)
• Forest Service-approved hardhat	• Forest Service-approved hardhat
• Eye protection	• Eye protection
• Appropriate gloves	• Appropriate gloves
• Heavy-duty, cut-resistant or leather, waterproof or water-repellent, 8-inch-high laced boots with nonskid soles	• Heavy-duty, cut-resistant or leather, waterproof or water-repellent, 8-inch-high laced boots with nonskid soles
• Hearing protection (85 decibels and higher)	• (Not required)
• Long-sleeved shirt	• (Optional)
• Chain saw chaps with a 2-inch boot overlap	• (Optional)

Personal protective equipment (PPE) should be used with engineering controls, substitution, administrative controls, or a combination of them to protect against hazards. Relying on PPE alone is not adequate.

General requirements should be followed for assessing the head, eye, face, hand, and foot hazards of a work project or activity.

General Requirements

- Select PPE based on hazards identified in the JHA.
    - PPE shall fit properly.
    - Defective, damaged, or unsanitary PPE shall not be used.
    - Supervisors shall assure the adequacy of PPE as well as its proper maintenance and sanitation.

- Each employee shall be trained to wear the PPE required by the JHA. Training shall include:
    - The required PPE and when and how it should be worn.
    - Proper care, maintenance, useful life, limitations, and disposal of PPE.

- Before performing any work project or activity requiring PPE, employees need to demonstrate an understanding of their training in its use. Employees are accountable for accidents and injuries that result from failing to use or misusing required PPE.

- Additional training may be necessary. Circumstances in which supervisors should provide additional training include:
  —Workplace changes that make earlier training obsolete.
  —Changes in the PPE to be used.
  —Evidence that an employee's knowledge or use of PPE is not adequate.

Specific Requirements

- Eye and Face Protection: Appropriate protection (including side protection) when employees are exposed to eye or face hazards such as flying particles, chemical gases or vapors, or potentially injurious light (such as ultraviolet light). Face shields can be used in saw operations in addition to safety glasses or safety goggles.

- Noise Protection: To comply with 29 CFR 1910.95, employees need to be in a hearing conservation program and wear ear plugs or ear muffs or both when working with equipment higher than 85 decibels (*Health and Safety Code Handbook* chapter 20, section 21.13b No. 2).

- Head Protection: All hardhats and helmets should be designed to provide protection from impact and penetration hazards from falling objects. Inspect shells daily for signs of dents, cracks, penetration, or any other damage that might compromise protection. Suspension systems, headbands, sweatbands, and any accessories should also be inspected daily.

- Hand Protection: Ensure that hand protection protects employees from the specific hazards that will be encountered. Gloves are often relied on to prevent cuts, abrasions, burns, and skin contact with chemicals that can cause local or systemic problems if they contact the skin (29 CFR 1910.138).

- Foot Protection: Footwear designed to prevent injury due to falling or rolling objects and objects piercing the soles. Heavy-duty, cut-resistant or leather, waterproof or water-repellent, 8-inch-high laced boots with nonskid soles are required for chain saw use.

- Additional Protection: Saw chaps, saw shoulder pads, or other PPE that provide cut resistance or puncture protection.

How Chain Saw Chaps Protect the User—When a chain saw strikes chain saw chaps, Kevlar fibers are pulled into the chain saw's drive sprocket, slowing and quickly stopping the chain.

A back-coated nylon shell covers the Kevlar protective pad inside the chaps. The shell resists water, oil, and abrasions. The protective pad consists of five layers of Kevlar in the following order: woven Kevlar, felted Kevlar, woven Kevlar, woven Kevlar, felted Kevlar. Kevlar is an aramid fiber similar to the Nomex material used in firefighter's clothing. Kevlar is more resistant to flame than Nomex. When chain saw chaps are exposed to temperatures higher than 500 degrees Fahrenheit, the nylon shell may melt, but the protective Kevlar pad will not burn.

Chain saw chaps need to be properly adjusted and worn snug to keep them positioned correctly on the legs. Chain saw users shall wear chaps. The chaps should provide coverage 2 inches below the boot tops. Proper fit and correct length maximize protection!

Chain Saw Chaps Specifications (MTDC-6170-4)—The Forest Service has provided cut-resistant protective chaps for chain saw sawyers since 1965. Chain saw chaps have prevented thousands of serious injuries.

The protective pad in the original Forest Service chain saw chaps consisted of four layers of ballistic nylon. Chain saw chaps tests conducted by the Missoula Technology and Development Center (MTDC) concluded that four layers of ballistic nylon offered protection to a chain speed of 1,800 feet per/minute (fpm) without a cut through. In 1981 Forest Service chain saw chaps were redesigned to improve the level of protection to a chain speed of 2,500 fpm without a cut through. The weight of the chaps was reduced by 40 percent, making them more comfortable.

The Center monitors chain saw injuries. Because chain saws require right-hand operation, the majority of chain contact injuries occur on the left leg. In 2000 the Forest Service chain saw chaps were redesigned. The new design provides protection to a chain speed of 3,200 fpm without a cut through and increases the area of coverage for the left side of the left leg by about $2\frac{1}{2}$ inches, and for the left side of the right leg by about $1\frac{1}{2}$ inches. The higher level of protection and larger area of protection increased the weight of each pair of chaps by 6 to 8 ounces, depending on the length (32, 36, or 40 inches). ***Only*** saw chaps provided by the General Services Administration meeting MTDC specifications 6170–4 are approved for purchase and use.

Inspection and Replacement—Chain saw chaps need to be inspected and replaced when appropriate. Replace chain saw chaps when:

- The outer shell has numerous holes and cuts. Holes in the outer shell allow bar oil to be deposited on the protective pad. The oil acts as an adhesive, preventing fibers in the pad from moving freely, decreasing the protection.

Chapter 1—Course Information and Safety Guidelines

- Wood chips and saw dust are evident in the bottom of the chaps.

- Repairs have stitched through the protective pad. Machine or hand stitching the protective pad prevents the fibers from moving freely, decreasing the protection.

- Cleaning has been improper. Detergents with bleach additives decrease the protection.

- High-pressure washing has destroyed the protective pad.

- The chaps have a cut in the first layer of yellow Kevlar that is more than 1 inch long.

Caring for Chain Saw Chaps—Treat your chain saw chaps as a ***CRITICAL*** piece of safety equipment. Keep them as clean as possible. Appropriate and timely cleaning reduces the flammability of the chaps and keeps your clothing cleaner. ***Do not use your chaps as a chain stop***.

Use Citrosqueeze, a commercially available citrus-based cleaning product, to clean chain saw chaps. Citrosqueeze has been tested and approved by Dupont for cleaning Nomex and Kevlar. Do not machine wash or machine dry chain saw chaps.

Cleaning Chain Saw Chaps—Hose and brush off chain saw chaps to remove dirt. Citrosqueeze must be diluted before use.

- For light soiling, use a Citrosqueeze solution in a spray bottle, mixing 1 part Citrosqueeze concentrate to 10 parts water. Spray solution on the area to be cleaned and brush the solution into the chaps with a bristle brush. Wait one-half hour, thoroughly rinse the chaps with cold water, and allow them to air dry.

- For heavy petroleum contamination, soak chain saw chaps in Citrosqueeze solution for a minimum of 4 hours, overnight if possible. Brush the chaps with a bristle brush, rinse them thoroughly with cold water, and allow them to air dry. Many pairs of chain saw chaps can be cleaned in a single soak tank. Use 10 to 15 gallons of solution in a soak tank.

A United States manufacturer for Citrosqueeze is:
    Emco Industries
    No. 118–2930 Norman Strasse Rd.
    San Marcos, CA 92069
    Phone: 888–727–3230

Repairs—Clean all chaps before repairing them. Repair cuts and holes in the outer shell as soon as possible to prevent the protective Kevlar pad from becoming contaminated with bar oil and petroleum products.

When repairing damage to the chaps' nylon shell, use a commercially available product called Seam Grip. Seam Grip provides a flexible, waterproof, and abrasion-resistant patch that will prevent petroleum products from contaminating the protective Kevlar pad.

Remove chain saw chaps from service if they have a cut longer than 1 inch in the top layer of Kevlar.

To repair holes and tears in the nylon shell:

1. Cut a piece of notebook or printer paper that extends about 2 inches beyond the edge of the damage.
2. Slip the paper inside the hole or tear so the paper lies on top of the protective Kevlar pad.
3. Lay the chaps on a flat, level surface and press the nylon shell down onto the piece of paper.
4. Squeeze Seam Grip onto the paper and onto the sides of the tear so that there is good coverage on all sides of the tear or hole.
5. Allow the patch to dry for at least 12 hours before using the chaps.

Seam Grip is available through outdoor retailers. To learn of retailers close to you, contact:
    McNett Corp.
    Box 996
    Bellingham, WA 98227
    Phone: 360–671–2227
    Fax: 360–671–4521
    Web site: http://www.mcnett.com

Situational Awareness

The situational awareness checklist can be used for self-assessment during sawing operations. It can also be used for discussions, tailgate safety sessions, or one-on-one problem solving (performance or skill deficiency) in the field.

Checklist of Personal Safety Considerations and Attitude

✔ How *do* I feel about this sawing assignment?

✔ Am I exercising sound judgment and awareness?

✔ Is my attitude influencing me to go against my better judgment (gut feeling)?

✔ Is my mind on my work project or activity?

✔ Do I have self-confidence?

✔ Am I overconfident?

✔ Am I doing this against my will? (*Health and Safety Code Handbook* chapter 20, section 22.48)

✔ Is peer pressure a factor?

✔ Am I professional enough to decline the assignment and ask for assistance?

✔ Do I have all of the required PPE and sawing equipment to do the job safely? Am I committed to using the PPE and equipment correctly?

✔ Am I complacent?

✔ Am I violating any safe operating procedures?

✔ Do I feel hurried or unusually stressed to get the tree on the ground or bucked?

✔ Have all options been considered and discussed with others?

✔ Am I in an unfamiliar environment and timber type?

✔ Do I watch out for my coworkers, contractors, and the public?

Evaluating the Complexity of the Assignment

The individual sawyer must determine the complexity of the assignment.

Your evaluation of the complexity of the assignment must be based on your individual skill, knowledge, and your understanding of your personal capabilities and limitations. The final decision to cut any tree is left up to the individual sawyer. You have the responsibility to say *no* and walk away from any sawing situation that is beyond your capabilities.

If a thorough job of assessing the complexity of the specific situation has been completed, the decision to cut or not to cut will be determined by the following Go, No-Go process.

Deciding Whether to Cut a Tree

Go! I feel comfortable with the sawing situation—I will cut the tree.

No Go! I don't feel comfortable with the situation—I will walk away from the tree.

Never base your decision on what you think you might be able to do. Remember…your safety and the safety of your coworkers depends on the decisions you make.

Glossary

This glossary is adapted from the *S–212 Wildfire Power Saws* training course.

Ax—A part of the faller's safety equipment used for pounding and chopping. It can also be used to plumb the lean of a tree.

Backcut—The last of the three cuts required to fell a tree. It is located on the opposite side of the tree from the undercut (face) and at least 2 inches (the stump shot) above the horizontal cut of the undercut (face). The backcut must never be continued to a point at which no holding wood remains.

Barber Chair—A tree that splits vertically when it is being felled. This is generally a result of improper facing or backcutting. A portion of the fallen tree is left on the stump.

Bind—A series of pressures in a felled tree resulting from objects (such as terrain or stumps) that prevent the tree from lying flat on the ground. The two major components of bind are compression and tension. Binds determine the technique and procedure used while bucking.

Blowdown—An area of timber blown over by strong winds or storms.

Bole—A tree stem thick enough for saw timber or large poles.

Boring—Using the nose or tip of the guide bar to saw into the tree while felling or bucking.

Bottom Bind—One of the four basic tree positions commonly encountered while bucking. A tree in a bottom bind is tensioned on top and compressed on the bottom.

Brushing—Removing the brush and shrubs while swamping out a work area.

Buck—Sawing through the bole of a tree after it has been felled.

Butt—The base of a tree stem.

Calks—Heavy boots containing numerous steel calks or spikes.

Conventional Undercut—The type of undercut commonly used to fell a tree. The undercut is taken from the butt of the tree.

Corners—The holding wood on either outside edge of the tree.

CPR—Cardiopulmonary resuscitation.

Danger Tree—A standing tree that presents a hazard due to conditions such as deterioration or physical damage to the root system, trunk, stem, or limbs, and the direction and lean of the tree.

Dogs (Bumper Spikes)—Chain saw accessory designed for felling and bucking. Medium-size saws will generally have an inside dog while larger saws will have an inside and an outside set of dogs. Chain saw dogs increase the sawyer's efficiency in felling and bucking operations.

Dolmar—Container for holding saw fuel and oil.

DOT—U.S. Department of Transportation.

Double Jack—A long-handled sledge hammer used to drive splitting and steel wedges.

Dutchman—A portion of the undercut that is not removed. A dutchman generally results when the horizontal and sloping undercut do not meet or extend beyond each other. A dutchman is very hazardous because it can change the felling direction.

End Bind—One of the four basic tree positions commonly encountered while bucking. An end bind situation occurs on steep terrain where the force of gravity closes the bucking cuts.

EPA—U.S. Environmental Protection Agency.

Escape Route—A predetermined path used by fallers when felling or bucking. Determine the direction and distance of the escape route and clear the route before cutting.

Face Cut—See undercut.

First-Aid Kit—A kit that includes bloodborne pathogen protective equipment (as a minimum, rubber gloves, face masks, eye protection, and CPR clear-mouth barriers) in addition to standard first-aid supplies.

Forest Service Approved—An item that meets Forest Service specifications or conforms to Forest Service drawings.

Guide Bar—The part of the chain saw that the saw chain travels on. Improper use of the bar (particularly the top and bottom of the bar at the end of the bar's nose) results in kickbacks and saw cuts.

Gunning (Sighting)—Aligning the handlebars or gunning mark with the desired felling direction. Because the handlebars or gunning mark are at a 90-degree angle to the bar, the exact position of the undercut can easily be established in relation to the desired felling direction. Some handlebars are not designed for gunning.

Hanging Wedge—A fan-shaped metal wedge.

Hangup—A situation in which a tree is lodged in another tree and does not fall to the ground.

Head Lean—One of the two natural leaning forces found in most trees. Head lean is more pronounced than side lean.

Holding Wood—Section of wood located between the undercut and the backcut. Its purpose is to prevent the tree from permanently slipping from the stump before it has been committed to the undercut. It also helps direct where the tree will fall. The holding wood must never be completely sawn off.

Hinge Wood—Same as holding wood.

Horizontal Undercut—The first of the two cuts required to undercut a tree. This level cut is at least one-third the diameter of the tree.

Humboldt Undercut—A type of undercut that is not recommended by the Forest Service for felling trees.

Itinerary—Planned route of travel, date of travel, destination, and estimated times of departure and arrival.

Jackstraw—Area where trees have been blown or fallen down in crisscross fashion.

JHA—Job hazard analysis.

Kerf—The slot saw-chain cutters make in the wood.

Kickback—A strong thrust of the saw back toward the faller generally resulting from improper use of the nose of the guide bar or from pinching the bar in a cut.

Lay—Refers to either the position in which a felled tree is lying or the intended location of a standing tree after it has been felled.

Lead—The established direction in which all trees in a quarter or strip are to be felled, usually governed by the terrain of the area, its general slope, or the skid road system.

Lean—The tilt of a tree away from its vertical position. Many times two leans may affect the same tree, such as head lean and side lean.

Leaner—A tree that leans heavily.

Limbing—Removing the branches from a felled or standing tree.

MSDS—Material safety data sheet. A compilation of information required under the Occupational Safety and Health Administration's Hazard Communication Standard that outlines the identity of hazardous chemicals, health, physical, and fire hazards, exposure limits, and storage and handling precautions.

NIOSH—National Institute on Occupational Safety and Health.

Offside—The opposite side of the tree from where the faller stands while bucking or felling.

OSHA—Occupational Safety and Health Administration.

Pie Shape (Wedge) Cut—A section sawn from a tree during bucking to allow for the directional pressures of various binds. Removing a pie-shaped section minimizes splits and slabs.

PPE—Personal protective equipment and clothing, respiratory devices, protective shields, and barriers.

Pistol-Grip Tree—A tree with a curve at the base of the trunk that makes it difficult to identify the tree's lean.

Safety Container—As defined by NFPA 30, an approved container with less than 5-gallon capacity, having a spring-closing lid and spout cover designed so that it safely relieves internal pressure when during a fire.

Sapwood—The outer layers of wood in growing trees that contain living cells and reserve material.

Side Bind—One of the four basic tree positions commonly encountered while bucking. A tree in a side bind is compressed on one side and tensioned on the other.

Side Lean—One of the two natural leans found in many trees. Side lean is less pronounced than head lean.

Sitback—Refers to a tree that settles back on the stump, closing the backcut's kerf. Sitback usually occurs because of improperly determining the tree's lean or by the wind.

Slabbing—A lateral split generally caused by improper technique or an improper sequence of bucking cuts.

Sloping Undercut—The second of the two cuts required to undercut a tree. This cut must be angled to allow a wide opening for the undercut.

Snag—Any standing dead tree.

Sound—Wood that is not rotten.

Spider—A gauge used for setting crosscut saw teeth.

Spike Top—A live tree that has a dead top.

Spring Pole—A limb or sapling that is bent under a tree or other weight.

Glossary

Stump Shot—Two inches or more height difference between the horizontal cut of the undercut (face) and the backcut. The difference in height establishes a step that will prevent a tree from jumping back over the stump toward the faller.

Swamp Out—To clean out brush and other material around the base of trees and where trees are to be bucked to protect against saw kickback and to provide safe footing.

Top Bind—One of the four basic tree positions commonly encountered while bucking. A tree in a top bind situation is compressed on top and tensioned on the bottom.

Undercut (Face Cut)—A section of wood sawn and removed from a tree's base. Its removal allows the tree to fall and helps direct where the tree will fall. The face is comprised of two separate cuts that have constant relationships. The horizontal cut must be at least one-third the diameter of the tree; the sloping cut must be angled enough to allow a wide opening, and the two cuts must not cross each other.

USDA—U.S. Department of Agriculture.

Wedge—A plastic or magnesium tool used by a faller to redistribute a tree's weight in the desired direction and to prevent a tree from falling backward. It is also used to prevent the guide bar from being pinched while bucking.

Widow Maker—A loose limb, top, or piece of bark that may fall on anyone working beneath it.

Sample Job Hazard Analysis

U.S. Department of Agriculture Forest Service	1. WORK PROJECT/ACTIVITY	2. LOCATION	3. UNIT
JOB HAZARD ANALYSIS (JHA) References: FSH 6709.11 and 6709.12 (Instructions on reverse)	4. NAME OF ANALYST	5. JOB TITLE	6. DATE PREPARED FS-6700-7 (03/00)
7. TASKS/PROCEDURES	8. HAZARDS	9. ABATEMENT ACTIONS (Engineering controls • substitution • administrative controls • PPE)	
Chain Saw Operation		**Qualifications** • Current certification by a nationally recognized organization to render first aid and perform CPR. Participation in an approved crosscut/chain saw program (Classroom and field training encompassing in part or in total a national training program, such as Wildfire Power Saws S-212). Supervisors—Ensure that saw operators receive training or retraining in first aid and CPR before certifications expire. Elements include: • Demonstration of sawing ability (to a certified operator or certified instructor) in functional areas. • Supervision by a certified instructor or certified operator of saw work by new operators. Supervisors—Monitor proficiency of sawyers to recognize the need for recertification (additional training) in less than 3 years. **Personal Protective Equipment (PPE)** Employees—Maintain PPE in a clean and fully functional condition.	
	Falling objects Flying or spraying objects Noise Sharp or pointed objects	Required PPE: • Forest Service–approved hardhat. • Eye protection. • Hearing protection (85 dB and above). • Appropriate gloves (cut-resistant gloves for chain filing). • Long-sleeved shirt. • Chain saw chaps (Forest Service–approved, minimum of 2 inches boot overlap). • Heavy-duty, cut-resistant or leather, waterproof or water-repellent, 8-inch-high laced boots with nonskid soles (hard toes are optional). • Fire shelter (wildfire and prescribed-burn assignments).	
		Required chain saw features: • Throttle interlock. • Felling and bucking spikes for felling and bucking operations (full set of two). • Antivibration system. • Chain brake, fully functional. • Proper saw for the job, fully operational (full wraparound handle bar for felling operations is required, three-quarter handlebars are allowed for bucking and limbing only). • Proper bar length for the specific work project or activity. • Bow bars with top and bottom chain guards and stingers. • Chain, filed and maintained.	
	Ergonomics, fatigue	General equipment: • First-aid kit. • Fire extinguisher. • Chain saw wrench. • Chain file with handle and guard. • Approved safety container for fuel. • Chain and bar oil container, clearly marked. • Proper wedges for the specific work project or activity (wooden wedges are not permitted). • Single-bit ax or maul, 3 to 5 pounds	

Sample Job Hazard Analysis

U.S. Department of Agriculture Forest Service	1. WORK PROJECT/ACTIVITY	2. LOCATION	3. UNIT
JOB HAZARD ANALYSIS (JHA) References: FSH 6709.11 and 6709.12 (Instructions on reverse)	4. NAME OF ANALYST	5. JOB TITLE	6. DATE PREPARED
7. TASKS/PROCEDURES	8. HAZARDS	9. ABATEMENT ACTIONS (Engineering controls • substitution • administrative controls • PPE)	

7. TASKS/PROCEDURES	8. HAZARDS	9. ABATEMENT ACTIONS
Transporting the Saw	Darkness Walking	Safety Practices: • No felling at night. • Carry so the bar (teeth) point downhill and away from the body—cover the bar if carrying on your shoulder. Prevent injury from cutters, dogs, and muffler. • Shut down the saw when carrying farther than tree to tree, or when slippery surfaces or brush create additional hazards.
	Vehicle	• Activate the chain brake for shorter distances. • Do not carry saws or fuel (including empty fuel containers) in the passenger compartment. • Do not store fuel and food together.
Situational Awareness and Sizeup		Analyze the cutting area by considering: • Location of people, structures, powerlines, and other obstacles. • Roads and travel in the cutting area. • Topography and steep ground. • Nearby hazards such as trees, low-hanging and dead limbs, rocks, and brush. • Primary and secondary escape routes, safety zones, and alternates. • Wind direction and velocity such as steady versus gusting and/or changing directions. • Tree species, both live and dead. • Diameter and height of trees. • Soundness of tree (split, lightning struck, broken-off top, rot, deterioration or physical damage to the root system, trunk, stem, limbs, or bark). • Lean direction. • Limb distribution. • Widow makers. • Spiked top. • Burning top. • Moisture (rain, snow, or ice).
Chain Saw Operation	Slips, trips, and falls	**Primary and Secondary Escape Routes, Safety Zones, and Alternates** • Select and prepare the work area by clearing a primary escape path and an alternate path before starting the cut.
	Walking surfaces	• Walk out and thoroughly check the intended lay of the tree. • Plan the route from the stump to the safety zone, generally not less than 20 feet away; the farther the better.
	Falling objects	• If possible, stand behind another tree, preferably quartering back from the planned direction of fall. Wait and watch for at least 30 seconds after the tree hits the ground for branches and other broken tree parts to fall. The shielding tree should be sound and large enough to provide protection.
Bucking, Brushing, and Limbing	Kickback Bind Rolling logs	• Know where the tip of the bar is at all times. • Anticipate log tensions (binds) and compressions and plan mitigation. • Use wedges and/or the pie cut. Initiate the cut slowly to observe the bind.
	Tension	• Use caution when cutting limbs supporting the log off the ground. Do not saw from the downhill side. On steep ground, prevent bucked sections from rolling or sliding. Limb from the top of large logs. • Watch for and carefully reduce tension on saplings and limbs with a series of small cuts on the tensioned side.

FS-6700-7 (03/00)

Sample Job Hazard Analysis

U.S. Department of Agriculture Forest Service	1. WORK PROJECT/ACTIVITY	2. LOCATION	3. UNIT
JOB HAZARD ANALYSIS (JHA) References: FSH 6709.11 and 6709.12 (Instructions on reverse)	4. NAME OF ANALYST	5. JOB TITLE	6. DATE PREPARED

FS-6700-7 (03/00)

7. TASKS/PROCEDURES	8. HAZARDS	9. ABATEMENT ACTIONS (Engineering controls • substitution • administrative controls • PPE)
Felling	Human factors Other hazards (kickback, bind, rolling logs)	• Consider your mental and physical condition. • Saw from a safe standing height. Be alert and look up frequently. The undercut must be clean with an opening large enough to control the tree nearly to the ground. Do not use corner or side cuts in hollow trees unless adequate holding wood can be maintained. Give a warning shout before beginning the backcut. Give another warning shout just before the tree falls. Insert a wedge into the backcut as soon as possible. In small-diameter trees, wedge into a corner cut. Do not cut off all of the holding wood. As the tree commits to the undercut, watch the top as you get quickly way from the stump. If the tree moves in a direction that compromises the primary escape route, use the alternate route. Do not leave a partially cut tree without marking it and warning others. When situations are deemed unsafe, use alternate methods or cancel the task.
Handling Flammable and Combustible Liquids	Burns, flammability, and toxic fumes	**Safety Practices** • A hazard communication training program provides information related to general awareness, hazard chemical inventory, and MSDSs. • A hazardous-chemical inventory shall be maintained and shall be readily accessible to all employees. • Never handle hazardous chemicals that do not have an MSDS. An MSDS is required from the manufacturer/supplier of each chemical used onsite and shall be readily accessible to employees at all times. **Transportation** • All containers (safety cans, drums, tanks, or tank trucks) used for transporting hazardous materials must be correctly labeled or placarded to ensure quick identification of the materials in an emergency. **Dispensing** • General Safety—All handling and dispensing of flammable liquids shall be done in a well-ventilated area free of sources of ignition, with bonding between the dispensing equipment and the container being filled.
Working Around Poisonous Plants	Accidental contact	**Procedures** • Teach all employees who are subject to exposure, especially those known to be highly sensitive, to recognize poisonous plants. When possible, do not assign allergic employees to jobs that expose them to poisonous plants. • Provide and apply a skin protectant or barrier cream. Fasten pant legs securely over boot tops (adhesive tape may be necessary). • Wear gloves and keep them away from the face and other exposed parts of the body. Do not touch skin with hands, clothes, or equipment that may have contacted poisonous plants. • Whenever the skin contacts a poisonous plant or noxious weed, wash the area with cold water within 1 to 3 minutes or as soon as possible. Use liberal amounts of water to ensure that all poisonous oils are washed off. While working around poisonous plants, do not wash with soap and/or hot water because they can remove natural protective oils from your skin. • Destroy poisonous plants around improved areas. • Avoid the smoke of burning poisonous plants. Inhaling this smoke can cause fever, malaise, respiratory problems, and severe rash. • Upon returning from the field, use rubbing alcohol to cleanse skin that contacted poisonous plants. • Clean tools with citric-based solvent before storing (use appropriate gloves and adequate ventilation). • Avoid exposure through mishandling contaminated clothes. Wash contaminated clothing separately from other clothes in hot water and detergent.

Sample Job Hazard Analysis

U.S. Department of Agriculture Forest Service	1. WORK PROJECT/ACTIVITY	2. LOCATION	3. UNIT
JOB HAZARD ANALYSIS (JHA) References: FSH 6709.11 and 6709.12 (Instructions on reverse)	4. NAME OF ANALYST	5. JOB TITLE	6. DATE PREPARED

FS-6700-7 (03/00)

7. TASKS/PROCEDURES	8. HAZARDS	9. ABATEMENT ACTIONS (Engineering controls • substitution • administrative controls • PPE)
Working Around Insects	Ticks	**Safety Procedures** • Spray clothes with an insect repellant, which may provide an additional barrier against ticks. Repellants, such as diethyl metoloamide (DEET), do not kill ticks. Some sprays do contain permethrin, which kills ticks on contact. Always follow the manufacturer's application instructions for insect repellants and treatments. • Wear light-colored clothing that fits tightly at the wrists, ankles, and waist. Each outer garment should overlap the one above it. Cover trouser legs with high socks or boots and tuck shirttails inside trousers. • Search the body repeatedly (such as during rest periods and lunch), especially hairy regions and inside clothing, as ticks seldom attach themselves within the first few hours. • Remove ticks with fine-tipped tweezers or fingers. Grasp the tick as close as possible to the point of attachment and pull straight up, applying gentle pressure. Wash the skin with soap and water then cleanse with rubbing alcohol. Do not try to remove the tick by burning it with a match or covering it with chemical agents. If the head pulls off when the tick is being removed, or if the tick cannot be removed, seek medical attention. • Once the tick has been removed, place it in an empty container so it can be given to a physician if you experience a reaction. Record the dates of tick exposure and removal. An early warning sign to watch for is a large red spot on a tick bite. Reactions within 2 weeks include fever, chills, headache, joint and muscle ache, significant fatigue, and facial paralysis.

Evacuation Plan (see attached Emergency Evacuation Plan)

10. LINE OFFICER SIGNATURE	11. TITLE	12. DATE

Field Site
EMERGENCY EVACUATION PLAN

Work project/activity: _General saw use_
Location: _____
Legal description: _____

To prepare for an emergency that requires first aid and/or immediate evacuation of personnel due to serious injury, *the following information shall be available to all crewmembers*:

• Designated first-aid provider(s): at least one person on each crew should be designated to provide first aid.
• Communication procedures to follow in the even of an emergency.
• Means of communication during duty hours: Forest radio to contact forest fire dispatch.

Additional Information for Sawyers

Chain Saw and Crosscut Saw Documents

Oregon Maintenance and Safety Manual. Blount, Inc., Oregon Cutting Systems Division; 4909 SE. International Way; Portland, OR 97222–4679; (or) P.O. Box 22127, Portland, OR 97269–2127.

Falling and Bucking Training Standard and Fallers and Buckers' Handbook. Workers Compensation Board of British Columbia, Films and Posters Section; P.O. Box 5350; Vancouver, BC V6B5L5.

An Ax to Grind (9923–2833–MTDC). Missoula Technology and Development Center; Bldg. 1, Fort Missoula; Missoula, MT 59804–7294.

Chain Saw and Crosscut Saw Videos

An Ax to Grind (99–01–MTDC). Missoula Technology and Development Center; Bldg. 1, Fort Missoula; Missoula, MT 59804–7294.

Be Smart—Be Sharp—Be Safe. Blount, Inc., Oregon Cutting Systems Division; 4909 SE. International Way; Portland, OR 97222–4679; (or) P.O. Box 22127, Portland, OR 97269–2127.

Chain Saw and Crosscut Saw Presentations

Situational Awareness Exercise for Chain Saws and *Situational Awareness Exercise for Crosscut Saws*. PowerPoint presentations. Missoula Technology and Development Center; Bldg. 1, Fort Missoula; Missoula, MT 59804–7294.

Chain Saw Videos

Chain Maintenance Clinic: Oregon Cutting. Workers Compensation Board of British Columbia, Films and Posters Section; P.O. Box 5350; Vancouver, BC V6B5L5.

Principles of Safe, Correct, and Efficient Chain Saw Use in All Tree Felling Operations. D. Douglas Dent, Inc.; P.O. Box 1099; Prineville, OR 97754.

Chain Saw Presentations

S–212 Wildfire Power Saws. Slide presentation. This training course is scheduled for revision fall of 2001. National Interagency Fire Center; 3833 S. Development Ave.; Boise, ID 83705.

Crosscut Saw Documents

Crosscut Saw Manual (7771–2508–MTDC). Revised May 1988. Missoula Technology and Development Center; Bldg. 1, Fort Missoula; Missoula, MT 59804–7294.

Lightly on the Land: The SCA Trail-Building and Maintenance Manual. Birkby, Robert C., ISBN# 0–89886–491–7, The Mountaineers, Seattle, WA, 1996.

Now You're Logging. Griffiths, Bus, ISBN# 1–55017–072–4, Harbour Publishing, Madeira Park, BC Canada, 1992.

Handtools for Trail Work. (8823-2601-MTDC). Revised February 1997. Missoula Technology and Development Center; Bldg. 1, Fort Missoula; Missoula, MT 59804–7294.

Northeastern Loggers' Handbook. Simmons, Fred C., USDA Agricultural Handbook No. 6, Northeast Forest Experiment Station, January 1951.

Food and Agriculture Organization of the United Nations, Basic Technology in Forest Operations. FAO Forestry Paper #36, ISBN# 92–5–101260–1, Rome, 1982.

Saws and Sawmills for Planters and Growers. Morris, John, ISBN# 1–871315–11–5, Cranfield Press, Bedford, UK, 1991

Logging Principles & Practices in the U.S. and Canada. Brown, Nelson Courtland, John Wiley & Son, Inc., 1934.

Country Woodcraft. Langsner, Drew, ISBN# 0–87857–200–7, Rodale Press, Emmaus, PA, 1978.

Crosscut Saw Reflections in the Pacific Northwest. Deaton, Jim, ISBN# 0–87770–675–1, Ye Galleon Press, Fairfield, WA, 1998.

Crosscut Saw Videos

Handtools for Trail Work (98–04–MTDC). Missoula Technology and Development Center; Bldg. 1, Fort Missoula; Missoula, MT 59804–7294.

Chapter 2—Chain Saw Use and Maintenance (Suggested time: 2 hours)

In this chapter, students will learn the field maintenance tasks for a chain saw, including:

- Removing the bar and the chain, inspecting them for damage and wear, and cleaning them.
- Removing and cleaning (or replacing) the air filter.
- Inspecting the power head for loose bolts and damage.
- Replacing the bar and the chain.
- Filing the chain.

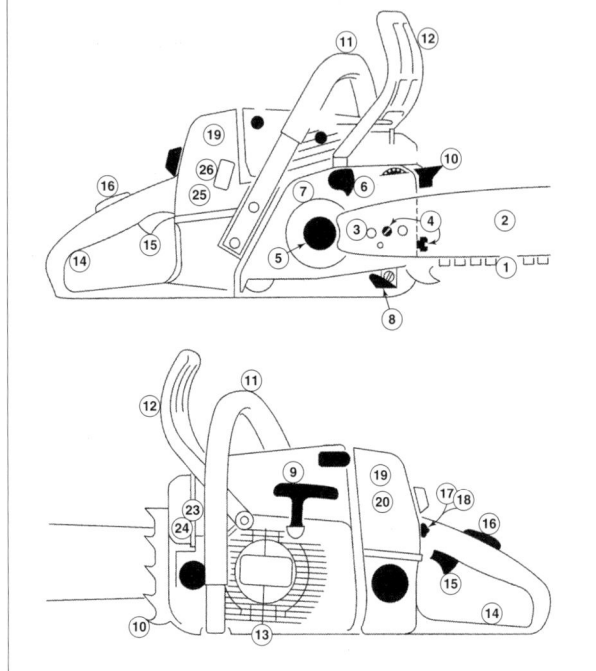

Chain Saw Operation

The bar and chain are the most important parts of your chain saw. A sharp chain produces shavings that fall to the ground away from the power head. A clean bar in good condition guides the chain through the cut, making a straight, true cut.

A dull chain produces sawdust that gets sucked into the air filter, cutting down the airflow to the power head and reducing power. A dull chain does not allow the saw to cut smoothly and puts unnecessary strain on the power head. The sawyer is forced to saw into the cut, increasing the stress on the power head. The improperly maintained bar and chain will damage the power head.

As the sawyer works harder to make the saw cut, the sawyer may become fatigued, increasing the risk of accident or injury. A dull chain also increases the risk of kickback.

The primary purpose of the chain saw and crosscut saw training and certification program is to provide for the safety of all employees who operate chain saws. Selecting the proper chain is important to operate chain saw safely.

1 Saw chain
2 Guide bar
3 Bar studs
4 Front and side chain tensioners
5 Chain sprocket
6 Chain brake
7 Clutch
8 Chain catcher
9 Starter grip
10 Bumper spikes (dogs)
11 Handlebar
12 Hand guard
13 Gunning sights
14 Rear handle
15 Throttle trigger
16 Throttle interlock
17 On/off switch
18 Choke
19 Air filter cover
20 Air filter
21 Fuel filter
22 Oil and fuel caps
23 Muffler
24 Spark arrester
25 Spark plug
26 Carburetor adjustments

—From Chain Saw Safety Manual, courtesy of Stihl, Inc.

Chain Saw Components

Saw chain—The three most common types of saw chains used by the Forest Service are chipper, chisel, and semichisel. Saw chain is made up of several parts that work together and must be properly maintained for maximum performance and safety.

The cutter is the part of the saw chain that does the cutting. The saw chain has left- and right-hand cutters so that the saw chain will cut evenly through the wood.

The depth gauge (referred to as a raker in some parts of the country) determines the depth of the cut (figure 2-1).

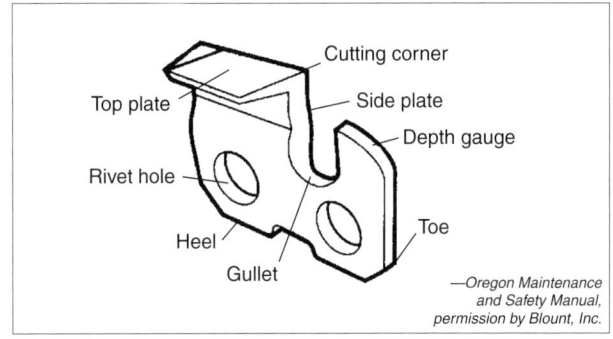

—Oregon Maintenance and Safety Manual, permission by Blount, Inc.

Figure 2-1—The depth gauge (raker) is the part of the tooth used to set the thickness of the shaving.

14

Chapter 2—Chain Saw Use and Maintenance

The three basic types of cutters include:

Chipper: The most versatile cutter type. Chipper (figure 2-2) chain is the easiest to file and will tolerate the most dirt and dust. Chipper chain cuts smoothly and is well suited for many Forest Service chain saw operations.

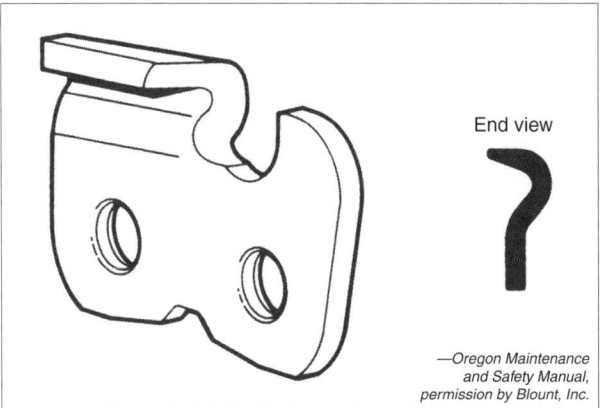

Figure 2-2—The chipper is the most versatile cutter type. Chipper chain is easy to file, will tolerate the most dirt, and can be used for many types of sawing operations.

Chisel (figure 2-3): The most aggressive cutter type. It is designed for production timber felling and should only be used by experienced sawyers. Chisel chain requires a file that fits the square shape of the cutting edge. It is more difficult to file than other types of chain. No file guide is available. Chisel chain dulls very quickly when it is exposed to dirt or dust. It is not recommended for brushing or limbing because of the potential for kickback.

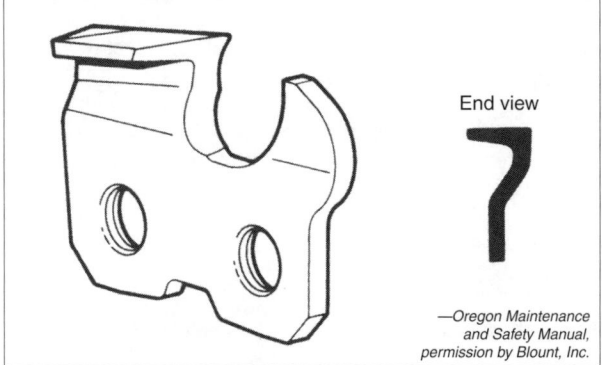

Figure 2-3—The chisel cutter is the most aggressive cutter type. chipper chain should only be used by experienced sawyers.

Semichisel (figure 2-4): A less aggressive cutter type than chisel. A round file is used with a file guide when filing semichisel chain. The semichisel cutter is more tolerant of dirt and dust and stays sharp longer than the other cutters.

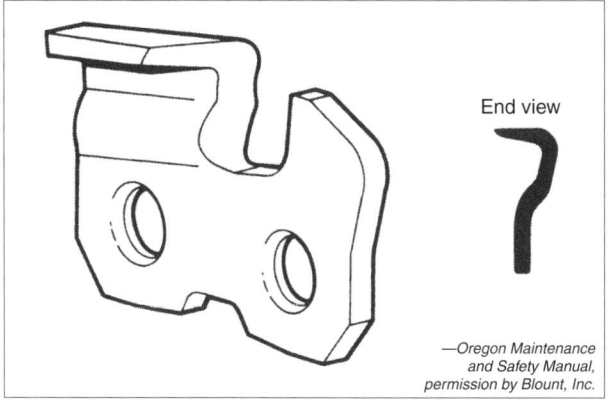

Figure 2-4—The semichisel cutter is less aggressive than the chisel cutter, will tolerate some dirt, and stays sharper than the chisel and the chipper cutters.

The low-kickback chain is the most desirable chain for training inexperienced sawyers. The chain cuts smoothly and is ideal for cutting brush, small-diameter material, dimensional lumber, house logs, and other materials that aren't normally cut with chain saws. Low kickback chain is available with chipper, chisel, and semichisel cutters.

Other chain parts (figure 2-5).

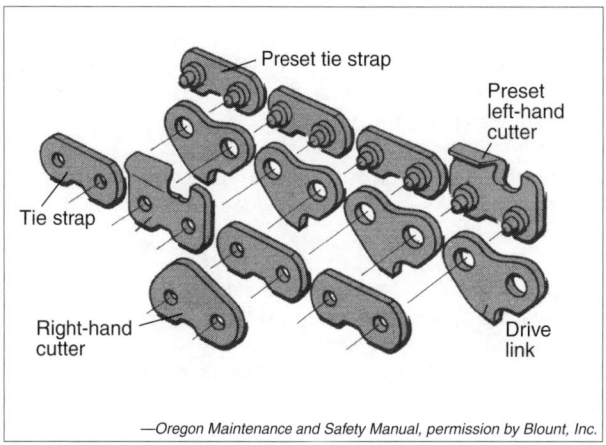

Figure 2-5—Parts of the chain.

15

Chapter 2—Chain Saw Use and Maintenance

Tie strap: Holds the parts of the saw chain together.

Drive link: Fits in the bar groove so the bar can guide the chain, and into the chain sprocket so the power head can drive the chain around the bar.

Cutter sequence (figure 2-6).

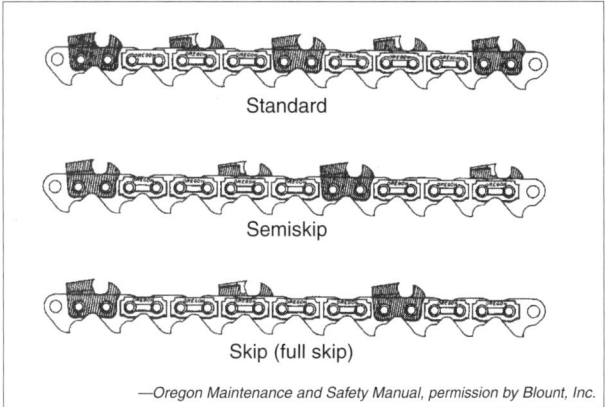

—Oregon Maintenance and Safety Manual, permission by Blount, Inc.

Figure 2-6—The cutter sequences for three types of chains: standard, semiskip, and skip (full skip).

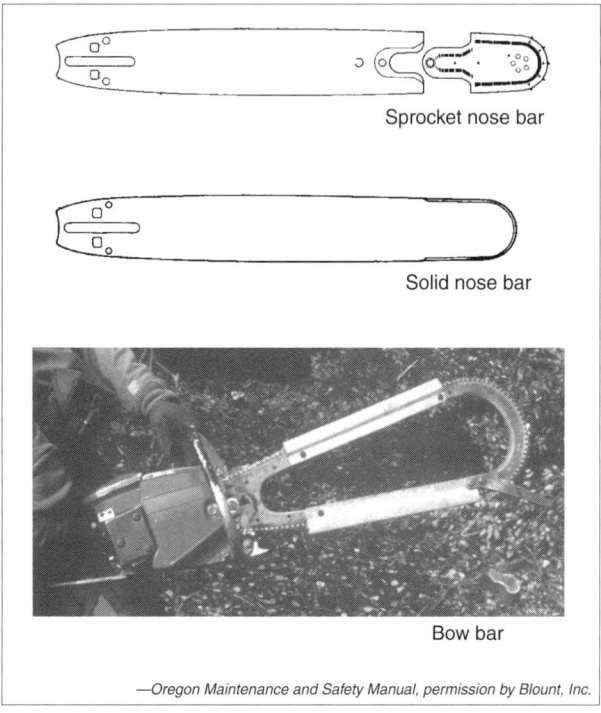

—Oregon Maintenance and Safety Manual, permission by Blount, Inc.

Figure 2-7—The three most common types of guide bars.

Standard: This chain has a cutter sequence of: left-hand cutter, tie strap, right-hand cutter, tie strap, left-hand cutter, tie strap, right-hand cutter, for the length of the chain.

Semiskip: This chain has a cutter sequence of: left-hand cutter, two tie straps, right-hand cutter, one tie strap, left-hand cutter, two tie straps, right-hand cutter, one tie strap, left-hand cutter, for the length of the chain.

Skip or full skip: This chain has a cutter sequence of: left-hand cutter, two tie straps, right-hand cutter, two tie straps, for the length of the chain.

Guide bar—The guide bar (figure 2-7) supports and guides the saw chain. The most common types of bars are solid nose, sprocket nose, and bow bar.

A sprocket nose bar has a sprocket in the nose to reduce drag and help the chain move freely around the bar.

A solid nose bar is generally found on small saws. The bar is solid with out a sprocket.

Bow bars form a large loop for the chain to follow. They are open in the center and are used most often for brushing. They have top and bottom chain guards and a stringer that the material being cut rests against.

Bar studs—Hold the bar and chain sprocket cover in place.

Front and side chain tensioner—Moves the guide bar to maintain proper tension on the saw chain.

Chain sprocket—Is the toothed wheel that drives the saw chain.

Chain brake—Stops the saw chain if it is activated by the sawyer's hand or by inertia (during kickback).

Clutch—Couples the engine to the chain sprocket when the engine is accelerated above idle speed.

Chain catcher—Helps reduce the risk of the saw chain contacting the sawyer if the chain breaks or if the chain is thrown off the bar.

Starter grip—A rubber or plastic handle attached to the starter pull rope.

Bumper spikes (dogs)—Hold the saw steady against wood.

Handlebar—Is used to hold the front of the saw.

Hand guard—Activates the chain brake and prevents contact with the chain if the sawyer's hand slips off the handlebar.

Chapter 2—Chain Saw Use and Maintenance

Gunning sights—Used to determine the planned direction of the tree's fall based on the undercut.

Rear handle—Used to hold the rear of the saw.

Throttle trigger—Controls the speed of the engine.

Throttle interlock—Prevents the throttle from being activated unless it is depressed.

On/off switch—Turns the saw on and off.

Choke—Used for starting a cold saw.

Air filter cover—Holds the air filter in place and covers the carburetor.

Air filter—Prevents dirt, dust, and sawdust from entering the carburetor.

Fuel filter—Prevents dirt and other contaminants from entering the saw's carburetor.

Oil and fuel caps—Seal the oil and fuel tanks.

Muffler—Reduces exhaust noise.

Spark arrester—Prevents hot sparks from leaving the muffler.

Spark plug—Ignites fuel in the power head.

Carburetor adjustments—Chain saws have a two-stage carburetor that provides fuel to the engine in any position that a saw may be held. The carburetor has three adjustments:

• Idle speed sets the speed at which the saw's engine will run by itself.

• Low-end speed controls the mixture of air and fuel on the first half of the throttle.

• High-end speed controls the mixture of air and fuel on the second half of the throttle.

The high- and low-end adjustments should be made by a qualified saw mechanic. Improper adjustment can result in poor operation or severely damage the chain saw.

The idle adjustment may need to be adjusted in the field. Before adjusting the idle, be sure that the air filter and fuel filter are clean and that you are using the right fuel mixture. Dirty filters or improper fuel mixtures affect the idle speed.

Newer saws designed to meet the Environmental Protection Agency (EPA) air quality standards may not have all three adjustments.

Mounts or antivibration system—Buffers between the engine and the handles that reduce vibrations to the sawyer's hands.

Guide Bar Maintenance

Most guide bar problems develop in the bar rails and are caused by:

• Incorrect chain tension
• Lack of lubrication
• Improper cutting techniques
• Normal wear

Look for several rail conditions (figure 2-8) when performing daily maintenance on your saw. These conditions can be corrected if they are caught early. If they are ignored, they will destroy the bar or lead to cutting problems. It may not be possible to cut straight or match cuts on larger material. In addition, the chain may be thrown because the chain tension is harder to control.

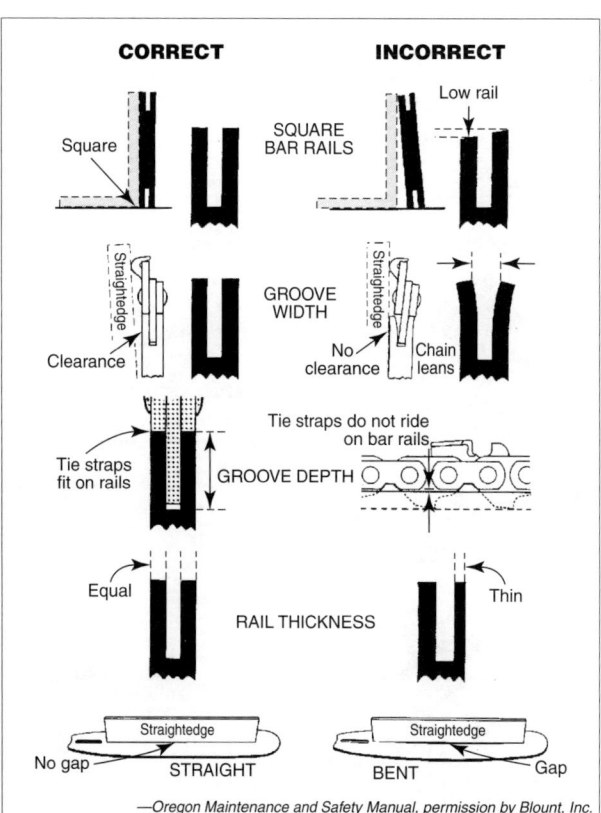

—Oregon Maintenance and Safety Manual, permission by Blount, Inc.

Figure 2-8—Correct rail conditions will prevent damage to the bar and cutting problems.

Chapter 2—Chain Saw Use and Maintenance

- Rails are worn down and the groove becomes shallow. If the groove is too shallow and the tie straps do not touch the rails, replace the bar.

- The outside edges of the rails develop wire edges. Use a flat file to remove them.

- The rail is worn low on one side. This causes the chain to cut at an angle. The bar will have to be ground on a specialized bar grinder. You may need to take the bar to a dealer or to a trained saw mechanic if your unit doesn't have a specialized bar grinder.

- The rails show blue discoloration along the bar or at the tip of the sprocket nose. This discoloration is caused by lack of lubrication, by poor cutting methods that push the drive links to the side, by a chain that is too tight, or by a dull or improperly filed chain. Blue spots are caused by excess heat. The spots are soft and will wear rapidly: you will need to replace the bar.

- The bar shows excessive wear only behind the nose on solid nose bars or behind the sprocket on sprocket nose bars. This wear can be caused by heavy use near the nose of the bar (such as limbing) or by a chain that is too loose. You can reduce this wear by periodically turning the bar over. If wear becomes extensive, you may need to replace the bar.

- The bar is bent. This can be caused by cutting techniques, getting the saw pinched or bound in the cut, and improper transportation, (such as carrying a saw loose in the bed of a pickup). Some bars can be straightened by a shop with the proper equipment.

The condition of the guide bar has as much to do with the performance of your chain saw as the condition of the chain. The bar and the chain work together. When both are in proper condition, the chain saw does the work. All you have to do is guide it.

Chain Tension

Remember three basic rules before tensioning a saw chain:

- Turn the saw off!
- Wear protective gloves.
- Wait until the bar and chain have cooled before adjusting the tension.

Heat causes the bar and chain to expand when the chain saw is being used. If the tension is set while the chain is hot, the chain will be too tight when it cools. Tension that is set too tight can damage the bar and chain.

To adjust the chain tension on a bow bar or solid nose:

- Loosen the bar nuts on the side of the saw.

- Pull the nose of the bar up and keep the nose up as you adjust the tension.

- Turn your saw's adjustment screw until the bottoms of the lowest tie straps and cutters just touch the bottom of the bar.

- Still holding the nose up, tighten the rear bar nut, then the front bar nut.

- Pull the chain by hand along the top of the bar several times from the engine to the tip. The chain should feel snug, but pull freely.

The tension must be tighter on a sprocket nose bar than on a solid nose bar. To adjust the tension on a sprocket nose bar:

- Loosen bar nuts on the side of the saw.

- Pull the nose of the bar up and keep the nose up as you adjust the tension.

- Turn your saw's adjustment screw until the bottoms of the lowest tie straps and cutters solidly contact the bottom of the bar.

- Pull the chain by hand along the top of the bar several times from the engine to the tip. The chain should feel snug, but still pull freely.

Daily Saw Maintenance

As the chain goes around the bar, it wears the bar and the chain. Because the bar is made of softer metal, the bar wears more than the chain. Generally, one rail will wear more than the other, causing the saw to cut at an angle if the bar and the chain are not properly maintained.

Chain saws have a chain oiler to minimize wear and prolong the life of the bar and chain. The oiler provides oil through a small hole in the bar that lines up with the oiler on the power head.

As oil is pumped through the oil hole, the chain carries it around the bar, lubricating the top, bottom, and roller tip. During operation, debris begins to build up in the chain groove. If the groove is not cleaned, oil cannot lubricate the entire bar, causing excessive wear and damage. If the oiler is properly adjusted,

a full tank of gas will run dry before the oil tank is empty. As a general rule, a tank of oil should last as long or longer than a tank of gas.

Clean and rotate the bar each time you file the chain or at least once a day. Be sure to clean the bar after filing the chain because fillings act as an abrasive, increasing the wear on the bar.

Cleaning Exercise

1. Remove the bar and chain for inspection and cleaning.
    • Check the bar for wear. Look for uneven rails, flared edges, cracks, and other damage that would require the bar to be repaired or serviced.
    • Clean the chain groove and oil holes. The proper method for cleaning the chain groove is to start at the tip with the bar tool and clean toward the base, moving debris away from the roller tip. Be sure that the oil holes are clean.
    • The sprocket nose (roller tip) should spin freely.
    • Grease the roller tip.
    • Bow bars require top and bottom chain guards and stingers. The guards may have to be removed before cleaning and filing the bar.

2. Remove and clean the air filter. Never use an air hose to blow out the air filter.
    • Take care not to damage the filter. Gently tap the filter against another surface. Don't rub or scrape it. Do not clean the filter with saw fuel. A damaged air filter can allow dust and debris into the engine, causing excessive wear and other problems.
    • Follow the manufacturer's recommendations (found in the instruction manual) for cleaning the air filter and determining whether it needs to be replaced. A dirty or plugged air filter reduces the power and performance and may cause other seemingly unrelated problems.

3. Check the muffler and spark arrester.

4. Remove the spark plug. Check for fouling. The tip of the plug should be beige, not black. The plug should be dry. Check the plug weekly when the saw is in frequent use.

5. Inspect the power head for loose bolts and damage. Tighten the bolts or repair the power head if necessary.
    • Check the handlebars for loose bolts or cracks.
    • Check the bumper spikes (dogs) for loose or bent bolts.
    • Check the mounts. They are the antivibration system. Look for cracks in the rubber. Excessive movement of the engine or a loose feeling when the saw is held by the handles and shaken indicates that the mounts may be broken or that they need tightening.

6. Replace the bar and chain.
    • Rotate the bar so that it wears evenly.
    • Check for proper alignment of the bar with the bar studs, adjuster, and oiler.
    • Check the chain tension. The chain should be adjusted so that it doesn't hang from the bar but still turns freely.
    • Check the chain brake to make sure it's operating properly.

Chain Maintenance

Chain maintenance is crucial to the performance of any chain saw. Before beginning any work assignment, follow four basic rules to maintain the saw chain for top performance and safe operation.

1. Your chain must be sharp. When your chain is sharp, the chain does the work. When the chain is dull, you do the work, making you fatigued and increasing the wear on the bar and chain.

2. Your depth gauges must be set correctly. The gauges' depths and shapes are critical to the saw's performance and your safety.

3. Your chain must be correctly tensioned. More bar and chain problems are caused by incorrect chain tension than by any other single condition.

4. Your chain must be well lubricated. Your bar, chain, and roller tip need a steady supply of oil. Otherwise, your bar and chain will be subject to excessive wear and damage.

Several conditions can increase the chain's potential for kickback, the risk of throwing or breaking the chain, or the risk of other hazards. Look for these conditions when inspecting your chain saw:

• Loose chain tension.
• Incorrect chain angles (generally caused by improper filing).
• Dull chain.
• Alteration of chain features designed to reduce kickback.
• Incorrect depth gauge settings (generally too deep).
• Improper shape of depth gauges after filing.
• Incorrectly installed chain parts.
• Loose rivets, or cracks and breaks in any chain part.

Chain Filing

This section focuses on chain filing with a round file and a clamp-on (handheld) file guide that clamps on the file, sometimes called a file holder. Using these files is the least complicated, least expensive, and most efficient way to file saw chain by hand in the field. Select a file that is the proper diameter for the saw chain.

After the saw chain has been hand filed a few times, it should be ground on a chain grinder to restore angles that may have changed during hand filing and to grind all cutters to the same length.

Understanding how a cutter works will help you see why proper chain maintenance is so important.

The depth gauge rides on the wood and controls the depth at which the cutting corner bites into the wood (figure 2-9).

The cutting corner and side plate sever the cross grains.

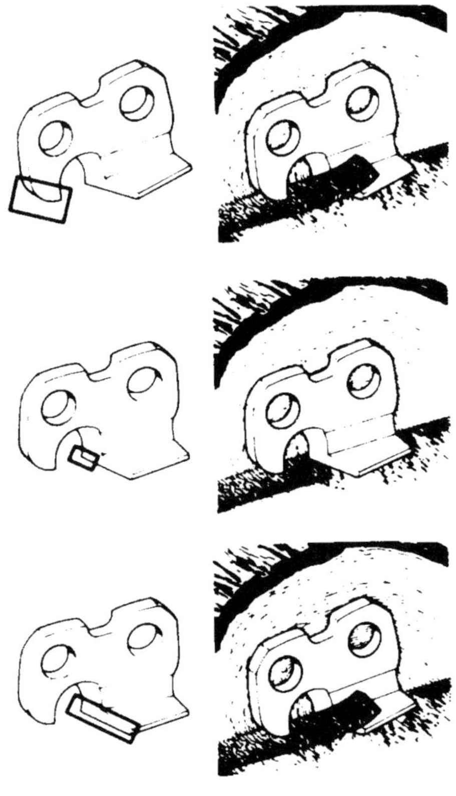

The top-plate cutting angle chisels out the severed wood fibers, lifting them from the kerf.

Three angles must be maintained when filing or grinding a saw chain (figure 2-10). A clamp-on file guide maintains these angles. The angles may vary on different types of saw chains.

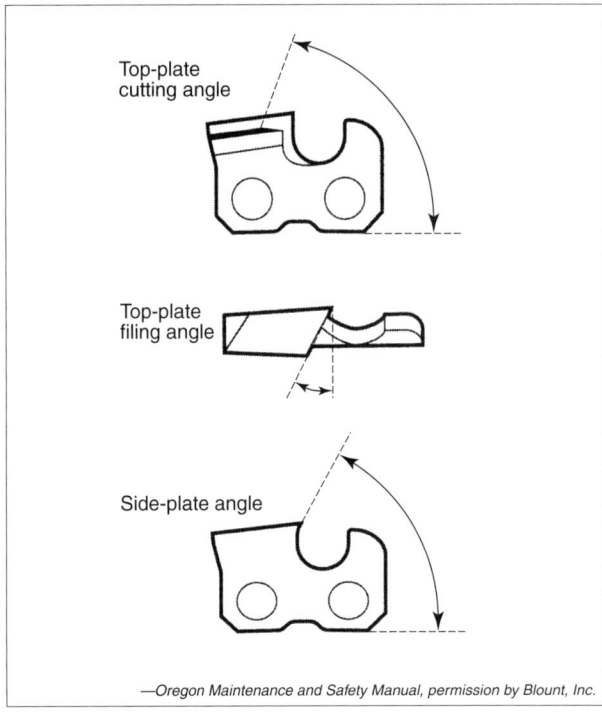

—*Oregon Maintenance and Safety Manual, permission by Blount, Inc.*

Figure 2-10—Maintain the top-plate cutting angle, top-plate filing angle, and side-plate angle.

Sharpening Cutters With a Round File

Be sure that the chain is tensioned properly. The file must be held at least one-fifth of the file's diameter above the cutter's top plate (figure 2-11). The clamp-on file guide positions the file for you.

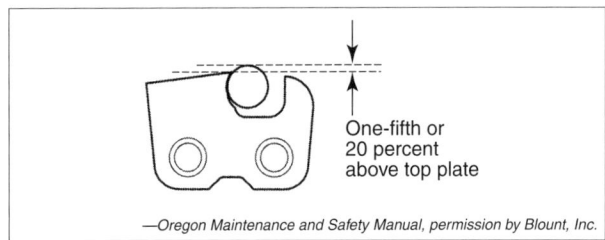

—*Oregon Maintenance and Safety Manual, permission by Blount, Inc.*

Figure 2-11—Hold the file at least one-fifth of the file's diameter above the cutter's top plate.

—*Oregon Maintenance and Safety Manual, permission by Blount, Inc.*

Figure 2-9—The depth gauge controls the depth at which a tooth's cutting corner bites into the wood.

Maintain the correct top plate angle (as marked on the file guide) by keeping filing angle parallel with your chain (figure 2-12).

Sharpen cutters on one side of the chain first, filing from the inside of each cutter to the outside. Turn the saw around and repeat the process for the remaining side (figure 2-13).

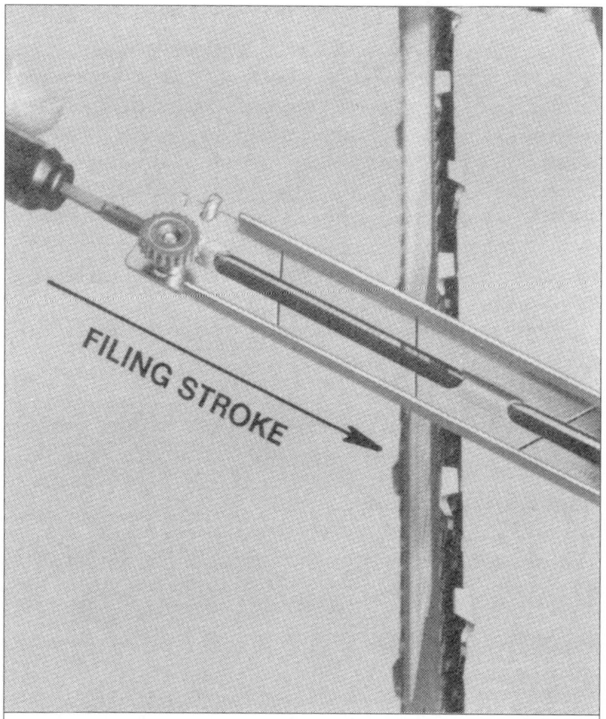

If the chrome surface of the top or side plates has been damaged, file until the chip has been removed from the chrome surface.

Keep the length of all cutters equal (figure 2-14).

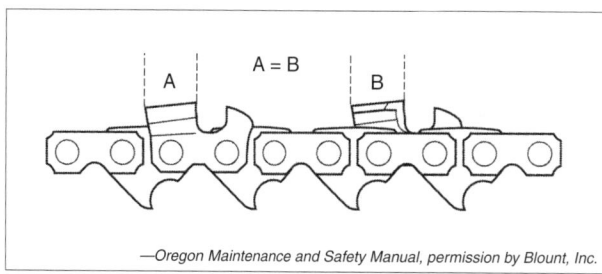

Figure 2-14—All cutters should be the same length.

How to Set Depth Gauges

Use a depth gauge tool with the correct built-in setting for the chain. Place the tool on top of the chain so one depth gauge protrudes through the slot in the tool (figure 2-15).

If the depth gauge extends above the slot, use a flat file to file the depth gauge level with the top of the tool. Never file a depth gauge lower than the top of the tool.

After lowering a depth gauge, round off its leading edge.

Figure 2-12—The correct top-plate angle is marked on the file guide.

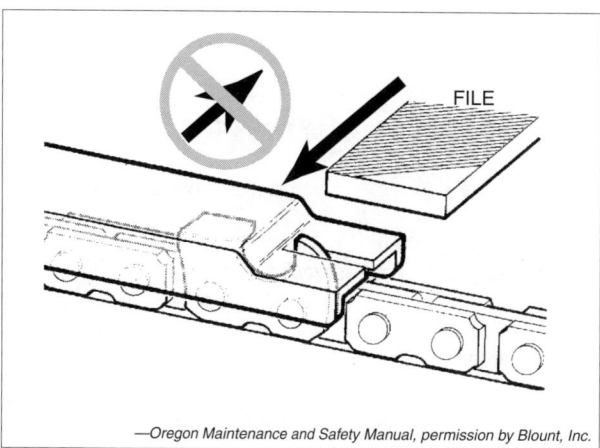

Figure 2-15—A depth gauge protrudes through the slot in the depth-gauge tool.

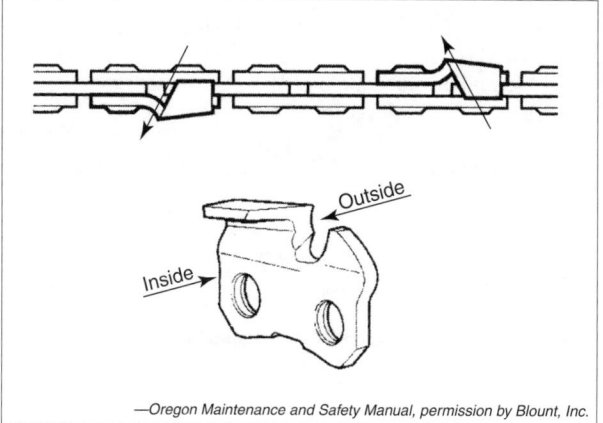

Figure 2-13—Sharpen the cutters on one side of the chain before turning the saw around to sharpen the other side.

Chain Filing Exercise

- Wear appropriate gloves for hand protection as described in your JHA.

- Make sure the chain is tensioned properly.

- Maintain the correct top-plate angle (as marked on the file guide) by keeping filing angle parallel with your chain.

- File one side of the chain, then the other.

- Keep the length of all cutters equal (figure 2-16).

- Set depth gauges with a depth-gauge tool.

—Oregon Maintenance and Safety Manual, permission by Blount, Inc.

Figure 2-16—After lowering a depth gauge, always round off its leading edge.

Saw Transportation

This section reviews three areas of saw handling: transporting the saw, starting the saw, and operating the saw.

Transporting Chain Saws in a Vehicle

- Keep the bar and chain covered with a chain guard.

- Properly secure the chain saw to prevent it from being damaged and to prevent fuel from spilling.

- Never transport a chain saw or fuel in a vehicle's passenger compartment.

Transporting Chain Saws by Hand

The muffler and power head can reach extremely high temperatures. Avoid these areas when carrying a saw that has been used recently.

- When carrying the saw for short distances, set the saw at idle speed and set the chain brake.

- When carrying the saw farther than from tree to tree, or in hazardous conditions (such as slippery surfaces or heavy underbrush), and in all cases if the saw is carried more than 50 feet, the saw shall be shut off and carried in a way that prevents contact with the chain, muffler, and bumper spikes (dogs).

- When carrying the saw on your shoulder, take extra care because of the sharpness of the chain and bumper spikes (dogs). A long-sleeved shirt, gloves, and a shoulder pad must be worn. The bar, chain, and bumper spikes (dogs) shall be covered, preferably with a manufactured bar and chain cover is recommended.

Safe Chain Saw Use

The methods to safely start and operate a saw can vary with the make and model.

- Maintain a secure grip on the saw at all times.
- Always start the saw with the chain brake engaged.
- Start the saw on the ground or where it is firmly supported.
- Do not "drop start" a chain saw.

Starting Procedures

Take extra care when starting your chain saw. Because you won't have both hands on the saw, you will need to be more careful to maintain complete control. Remember that on/off switches may vary with different makes of saws.

- Ensure that appropriate PPE is available and is worn correctly.

- Do not "drop start" a chain saw. This is the most dangerous method of starting a saw because you have no control of the saw.

- Make sure that the saw's bar and chain do not contact anything.

- Always start the saw with its chain brake engaged.

- Maintain a firm grip on the saw at all times.

- Start the saw on the ground or where it is firmly supported with the nose of the saw bar over a stump or log. To successfully start a cold saw, feather the throttle trigger, providing adequate throttle. Avoid engaging the throttle lock or the fast idle position of the on/off switch.

Starting the Chain Saw on the Ground

- Set the chain brake.

- Place the saw on firm ground in an open area (figure 2-17).

- Grip the front handlebar firmly with your other hand.

- Place the toe of your right foot into the rear handle and press down.

- Pull the starting rope with your other hand until you feel resistance.

- Give a strong, brisk pull.

—From Chain Saw Safety Manual, courtesy of Stihl, Inc.

Figure 2-17—Start the saw on firm ground.

Operational Safety

A full-wrap handlebar allows cutting from both sides of the tree using the bottom of the bar, the bar's most aggressive part. In some situations the ability to cut wood rapidly is critically important for safety. Full-wrap handlebars are designed to be used by both the left and the right hand. The sawyer's thumb should always be wrapped completely around the handlebar, no matter how the saw is turned. The thumb and fingers are essential for maintaining control of the chain saw, especially during a kickback. The grip on the chain saw should be firm, but not overly tight.

Handling

- Never operate a chain saw with one hand. You do not have control of the saw and increase the risk that you will be injured if the saw kicks back.

- Always grip the saw firmly with both hands, the dominant hand on the front handlebar and the other hand on the throttle and rear handle.

- Place your fingers tightly around the handle and the handlebar, keeping them between your thumb and forefinger.

- Never operate a chain saw with the throttle lock engaged. If you do, you cannot control the saw or the chain speed.

- Make sure your work area is clear of people and obstacles, such as rocks, stumps, holes, or roots that may cause you to stumble or fall.

- Make sure that the saw chain does not contact any materials such as rocks or wire. Such contact is a safety hazard and will dull the chain. The chain will require filing or it may be damaged in ways that filing cannot correct.

Reactive Forces

The laws of physics explain that for every action there is an equal and opposite reaction. These reactions happen very quickly during chain saw operation and can be dangerous.

Kickback—Kickback is the most powerful reactive force you will encounter while operating a chain saw. Kickback can occur while felling, limbing, bucking, or brushing when the upper quadrant of the bar nose contacts a solid object or is pinched (figure 2-18).

Chapter 2—Chain Saw Use and Maintenance

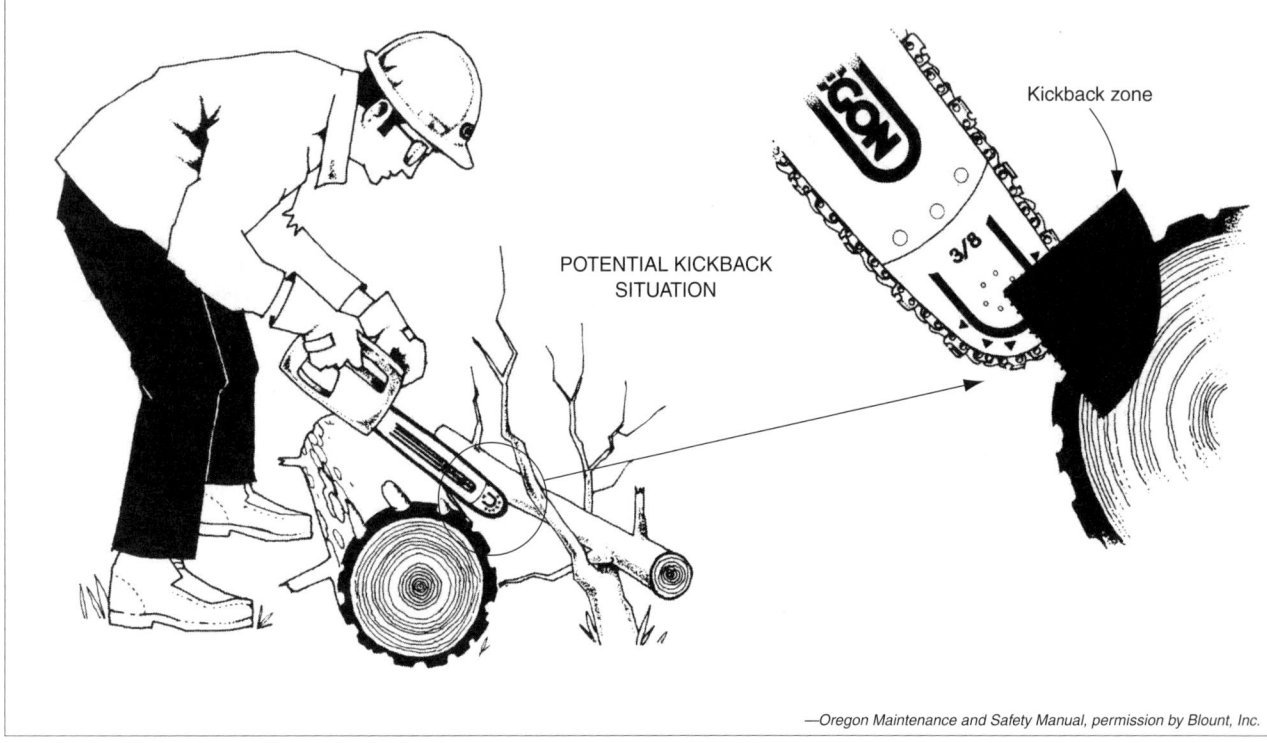

Figure 2-18—Kickback occurs when the upper quadrant of the bar nose contacts a solid object or is pinched.

During kickback, the bar is forced up and back in an uncontrolled arc toward the sawyer. Many factors determine the severity of the kickback and the arc, including: chain speed, angle of contact, condition of the chain, and the speed at which the bar contacts the object.

Ways to avoid kickback:
- Hold the saw with both hands, securely gripping the handle and the handlebar between your thumb and forefinger.
- Be aware of the location of the bar nose at all times.
- Never let the bar nose contact another object.
- Never cut with the power head higher than your shoulder.
- Never overreach.
- Pull the saw smoothly out of the cuts. This technique will help to reduce kickbacks and fatigue.
- Cut one log at a time.
- Stand to the side of the kickback arc.
- Use caution when entering a partially completed cut.
- Use a properly sharpened and tensioned chain at all times.
- Watch the cut and the log for any movement that may pinch the chain.
- Use a low kickback chain.

Pushback—Pushback occurs when the chain on the top of the bar is suddenly stopped by contacting another object or by being pinched. The chain drives the saw straight back toward the sawyer (figure 2-19).

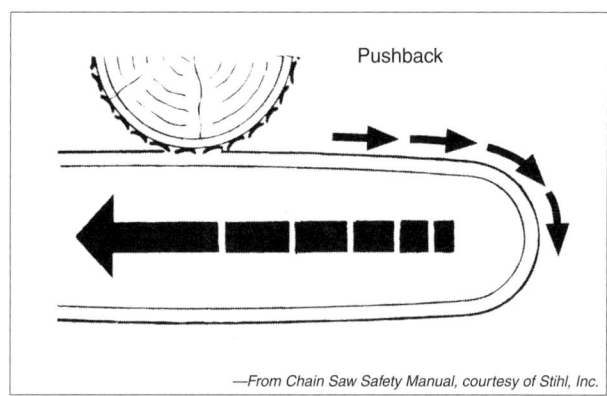

Figure 2-19—Pushback occurs when the chain on the top of the bar is suddenly stopped by contacting another object or by being pinched.

Chapter 2—Chain Saw Use and Maintenance

Ways to avoid pushback:
- Only cut with the top of the bar when necessary.
- Watch the cut and the log for any movement that may pinch the top of the bar.
- Do not twist the bar when removing it from a boring cut or underbuck.

Pull-In—Pull-in occurs when the chain on the bottom of the bar is caught or pinched, and suddenly stops. The chain pulls the saw forward (figure 2-20).

Ways to avoid pull-in:
- Always start a cut with the chain at or near full speed and the bumper spikes (dogs) contacting the wood.
- Watch the cut and the log for any movement that may pinch the bar. Use wedges to keep the cut open.

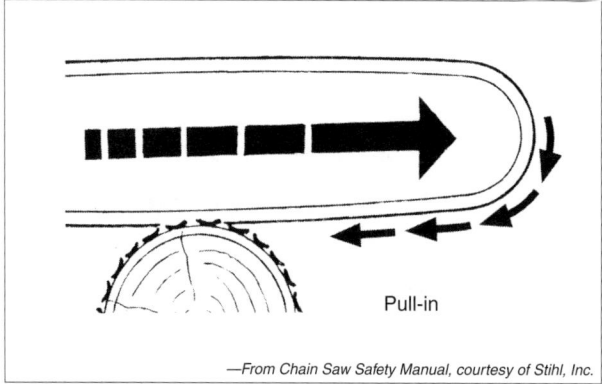

—From Chain Saw Safety Manual, courtesy of Stihl, Inc.

Figure 2-20—Pull-in occurs when the chain on the top of the bar is suddenly stopped by contacting another object or by being pinched.

Additional Tools

This section includes information about axes, wedges, approved safety containers for fuel and oil, peaveys, and cant hooks.

Axes

Axes are used to remove bark from trees and to drive wedges during felling and bucking. The ax handle should be smooth and free of cracks. The head should be securely attached to the handle. Axes used for driving wedges should have a straight handle.

Axes need to be heavy enough (3 to 5 pounds) to drive wedges into the trees being felled. The back of the ax should be smooth, have rounded edges, and be free of burrs to minimize damage to wedges. Pulaskis should never be used to drive wedges.

Always remove branches, underbrush, overhead obstructions, or debris that might interfere with limbing and chopping. Do not allow anyone to stand in the immediate area. Make sure workers know how far materials may fly. Protect all workers against flying chips and other chopping hazards by requiring them to wear the appropriate PPE.

Always position your body securely while working with a tool. Never chop crosshanded; always use a natural striking action. Be alert when working on hillsides or uneven ground. If you cut a sapling that is held down by a fallen log, the sapling may spring back. Be alert for sudden breakage. If you do not have a need to cut something, leave it alone.

Never use chopping tools as wedges or mauls. Do not allow two persons to chop or drive wedges together on the same tree. When chopping limbs from a felled tree, stand on the opposite side of the log from the limb being chopped and swing toward the top of the tree or branch. Do not allow the tool handle to drop below a plane that is parallel with the ground unless you are chopping on the side of a tree opposite your body.

If the cutting edge picks up a wood chip, stop. Remove the chip before continuing. To prevent blows from glancing, keep the striking angle of the tool head perpendicular to the tree trunk.

Wedges

Wedges are essential tools for safe felling and bucking. They provide a way to lift the tree, preventing the tree from sitting back when it is being felled. A wedge must be inserted into the backcut as soon as possible. Wedges also reduce binds on the saw when bucking.

Select the correct wedge for the job. The proper type, size, and length or a wedge varies, depending on its use. The size of the tree being felled or the material being bucked determines the size of the wedge that will be needed. If the wedge is too small, it may be ineffective. If the wedge is too long, it may not be able to do its job without being driven so far into the tree that it contacts the chain.

Always drive wedges by striking them squarely on the head. Drive them carefully to prevent them from flying out of the cut.

Check wedges daily or before each job. Do not use cracked or flawed wedges. Wedges that are damaged need to be cleaned up before they are used again.

Recondition heads and the tapered ends when grinding wedges to the manufacturer's original shape and angle. Wear eye protection and a dust mask.

Repair any driving tool or remove it from service when its head begins to chip or mushroom.

Carry wedges in an appropriate belt or other container, not in the pockets of clothing.

Most wedges are made out of plastic or soft metal, such as magnesium, and come in different sizes. Use plastic wedges in both felling and bucking operations to prevent damaging the saw chain if it contacts the wedges.

The two basic types of wedges used in sawing are single and double taper.

Single-taper wedges (figure 2-21) are simple inclined planes designed to provide lift during tree felling. As the wedge is driven into the back cut, the tree hinges on the holding wood, redistributing the tree's weight. The sawyer must coordinate striking the wedge with the forward sway of the tree, allowing the wedge to be driven more easily and sending less of a shock wave up the tree. Striking the wedge when the tree is in its backward sway sends a severe shock wave up the tree and can knock out dead branches or tops, endangering the sawyer. Sawyers should look up for falling material after each blow to the wedge.

Double-taper wedges (figure 2-22) are designed to reduce bind. They taper equally from the centerline, forcing the wood to move equally in both directions. They perform best when used in bucking to prevent the kerf from closing and binding the guide bar.

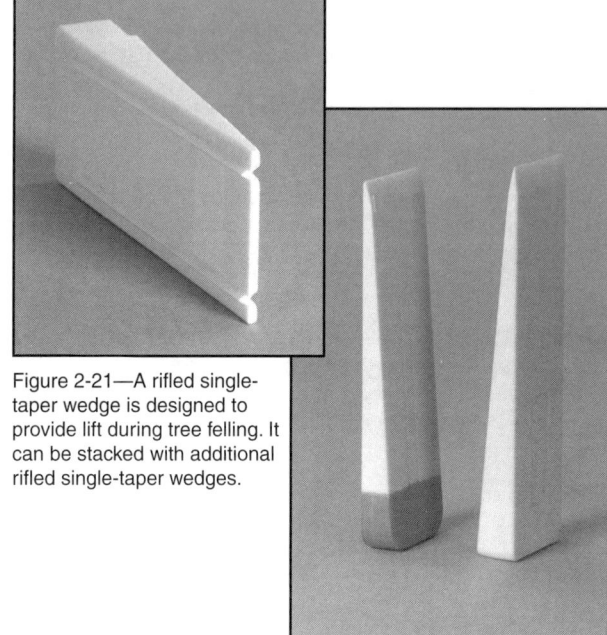

Figure 2-21—A rifled single-taper wedge is designed to provide lift during tree felling. It can be stacked with additional rifled single-taper wedges.

Figure 2-22—Double-taper wedges are designed to reduce bind.

Fuel and Oil Containers

The most commonly used fuel and oil container is the two-chambered (DOLMAR type) safety container. Transport the safety container with all lids fully sealed. Empty the container thoroughly before storage.

Even empty containers are dangerous. Large quantities of saw fuel need to be transported in an approved safety can.

- If a container is missing a lid or showing signs of a defect such as cracks, take it out of service immediately.

- All employees who handle, transport, or use flammable or combustible liquids shall receive hazard communication standards training and be familiar with material safety data sheets.

- Passengers shall not ride in the enclosed cargo portion of a vehicle hauling flammable or combustible liquids. If it is absolutely necessary to carry flammable or combustible liquids with a passenger vehicle, a minimum amount of such cargo shall be secured in a rack on the roof.

- Never transport fuel in the same cargo area with oxidizers, acids, or radio equipment.

- Flammable or combustible liquids shall be carried in approved safety containers as defined by the National Fire Protection Association (NFPA 30). Such containers shall be clearly labeled to identify the contents.

Containers shall never be filled more than 90 percent with fuel. Fuel vapors need room to expand. Because the two-chambered (DOLMAR type) safety container is not equipped with a spring-loaded lid, a chain saw should be fueled only after the saw has cooled completely.

- Allow the saw to cool for at least 5 minutes before refueling.
- Fill the saw on bare ground or other noncombustible surface.
- Immediately clean up spilled fuel.
- Refuel outdoors and at least 20 feet from any open flame or other sources of ignition.
- Do not start the saw closer than 10 feet from the fueling area.

Peaveys and Cant Hooks

The blacksmith Joseph Peavey invented the peavey. Both the peavey and the cant hook use a curved metal hook on the end of a straight handle to roll or skid logs. A peavey has a sharp pointed spike at the lower end, while a cant hook has a tow or lip. Most peaveys and cant hooks come with a duckbill hook that is a good all-around style. Peaveys and cant hooks come with hickory handles that are from 2 to 5½ feet long.

Peaveys are used almost exclusively in the woods where the pick is used to for prying. Peaveys are handy for prying logs up onto blocks to keep the saw from pinching while bucking. The cant hook is used primarily to roll logs.

- Keep the handle free of splinters, splits, and cracks.
- Keep the point sharp.
- Keep your body balanced when pushing or pulling the pole.
- Grip the handle firmly. Do not overstress it.
- Place a guard on the point when the tool is not in use.

Chapter 3—Chain Saw Tasks and Techniques (Suggested time: 2 hours)

In this chapter:

- Students will learn the importance of a thorough sizeup for felling, limbing, and bucking operations.
- Students will acquire the skills to operate a chain saw safely and efficiently.

Safe Chain Saw Use

Proper Use of Bumper Spikes (Dogs)

Learn to use the saw's bumper spikes (dogs) as a pivot point when felling or bucking. This technique will enhance your control of the saw and improve the saw's efficiency while reducing fatigue.

Cutting with the bottom of the bar pulls the chain saw away from the sawyer. Cutting with the top of the bar pushes the saw back at the sawyer. Cutting with the bottom of the bar increases efficiency and decreases the sawyer's fatigue.

Always protect the saw chain from becoming dull. This will reduce unnecessary fatigue and lessen chances of kickbacks and barber chairs. Barber chairs will be addressed later. Keep the chain out of the dirt and rocks. When cutting uprooted trees with dirt and rocks in the bark, use an ax to trim the bark away from the area to be cut.

Bucking

Situational Awareness

- Never buck a tree that exceeds your ability.
- Consider overhead hazards.
- Is the guide bar long enough for the tree that is being bucked?
- Establish good footing.
- Swamp out bucking areas and escape routes. Anticipate what will happen when the log is cut.
- Plan the bucking cut carefully after considering:
  —Slope: People and property in the cutting zone.
  —Tension: Spring poles.
  —Compression: Falling or rolling root wads.
  —Rocks and foreign objects on the log: The log's tendency to roll, slide, or bind.
  —Pivot points: Broken off limbs hidden underneath the log that can roll up and grab the sawyer.

The Forest Service *Health and Safety Code Handbook* emphasizes the following points.

- Assess the area for overhead hazards before beginning bucking.
- Size up the log for possible reaction after the release cut has been completed.
- Establish escape routes and clear any obstacles that might inhibit your escape.
- Cut slowly and observe the kerf for movement that will indicate where the bind is. A log can have different types of binds at different places.

Safe and Efficient Bucking Techniques

In most situations it is safest to buck logs from the uphill side unless the log may move uphill when bucked. This could occur because of the log's position, weight distribution, and pivot points. Always consider binds and pivot points. Consult another sawyer if you have questions.

Begin bucking by cutting the offside first. This is the side the log might move to when it is cut, normally the downhill side. Cut straight down until you have space for a wedge.

Insert a wedge or wedges to prevent the cut (kerf) from closing tightly and pinching your bar.

Understanding directional pressures—or binds—is important for safe and efficient cutting. These binds determine bucking techniques and procedures. Look for landforms, stumps, blowdown, and other obstacles that prevent a log from lying flat, causing binds. When a bind occurs, different pressure areas result. The tension area is the portion of the log where the wood fibers are being stretched apart. In this portion of the log, the chain saw's cut (kerf) opens as the cut is made. The other pressure area is called the compression area. Here the wood fibers push together. In this portion of the log, the kerf closes as the cut is made (figure 3-1).

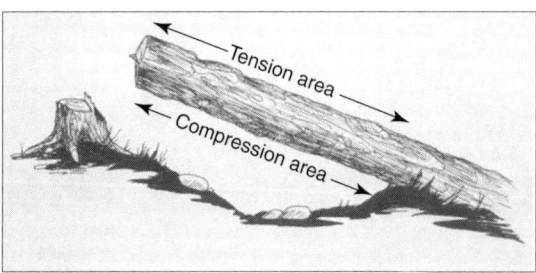

Figure 3-1—When a bind occurs, pressure areas result. These areas are called tension (pulling apart) and compression (pushing together) areas.

Chapter 3—Chain Saw Tasks and Techniques

It is extremely important to determine what will happen to the log when it is cut. Inspect the log for all binds, pivot points, and natural skids. Various bucking techniques can be used to lower a suspended tree to the ground.

Determining Bind

The four types of bind are: top, bottom, side, and end. There may be a combination of binds. Normally, logs have a combination of two or more binds (figure 3-2).

- Top bind—The tension area is on the bottom of the log. The compression area is on the top.

- Bottom bind—The tension area is on the top of the log. The compression area is on the bottom.

- Side bind—Pressure is exerted sideways on the log.

- End bind—Weight compresses the log's entire cross section.

It is best to start bucking at the top of the log and work toward the butt end, removing the binds in smaller material first. Look for broken limbs and tops above the working area. Never stand under an overhead hazard while bucking.

Look for small trees and limbs (spring poles) bent under the log being bucked. They may spring up as the log rolls away. If you can safely do so, cut these hazards before the log is bucked (figure 3-3). Otherwise, move to a new cutting location and flag the hazard.

Determine the offside. It is the side the log might move to when it is cut—normally the downhill side. Watch out for possible pivots. Clear the work area and escape paths. Allow more than 8 feet of room to escape when the final cut is made. Establish solid footing and remove debris that may hinder your escape (figure 3-4).

Cut the offside first. If possible, make a cut about one-third the diameter of log. This allows the sawyer to step back from the log on the final cut. Do not let the tip of the bar hit any object.

Watch the kerf to detect log movement. Position yourself so you can detect a slight opening or closing of the kerf. There is no better indicator of the log's reaction on the release cut. If the bind cannot be determined, proceed with caution. It may be necessary to move the saw back and forth slowly in the kerf

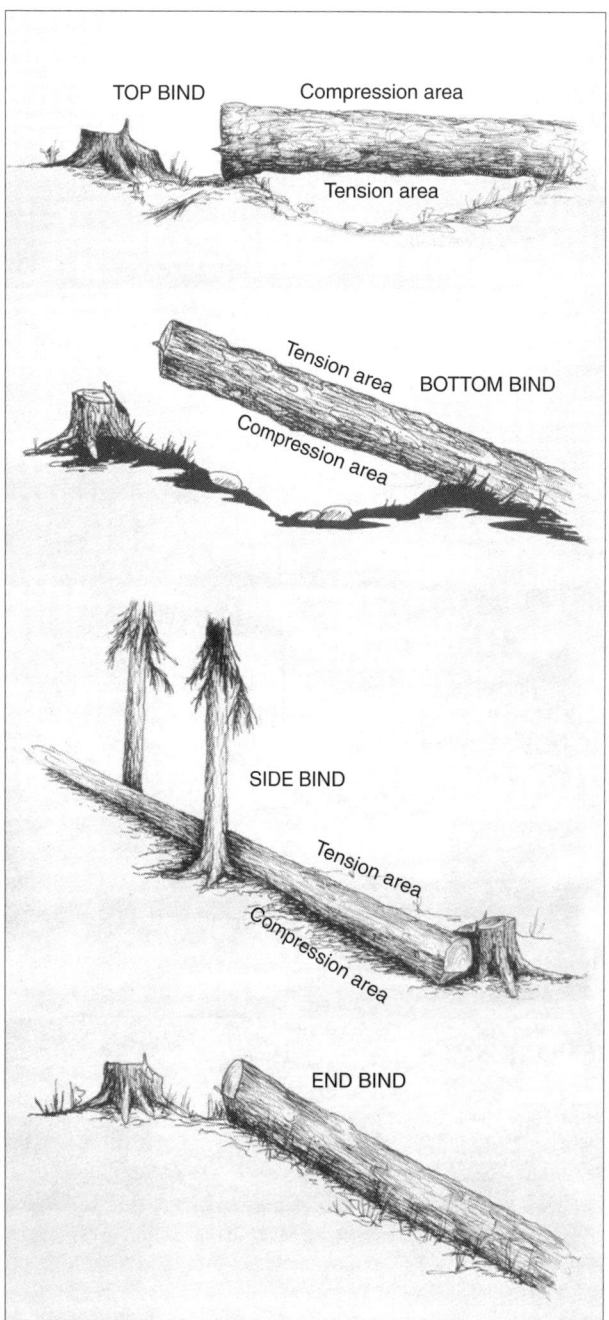

Figure 3-2–There are four types of binds. A log can have a combination of two or more binds.

Chapter 3—Chain Saw Tasks and Techniques

Figure 3-3—Look for spring poles. They can release and cause accidents.

Figure 3-4—Determine and stay clear of the offside (downhill side) when you are bucking.

Figure 3-5—Watch the kerf for movement that will indicate a bottom bind (kerf opens) or top bind (kerf closes).

Reduce remaining wood. Visually project the kerf's location to the bottom of the log. Reduce the amount of wood for the final cut by cutting a short distance into the log along this line. Be prepared for kickback.

Determine the cutting sequences (figure 3-6).

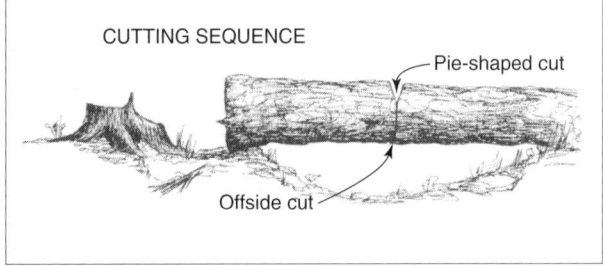

Figure 3-6—Determine the cutting sequence before beginning to cut.

(about 3 seconds for each chain revolution) to prevent the saw from getting bound as the kerf closes behind the guide bar. Cut only deep enough to place a wedge. Continue cutting. Watch the kerf (figure 3-5). If the kerf starts to open, there is a bottom bind; if the kerf starts to close, there is a top bind.

The sequence of the remaining cuts depends on the type of bind. Generally the next cut will be a small, less than 1½ inches, pie-shaped cut removed from the compression area. The log can settle slowly into this space, preventing dangerous slabbing and splintering. This practice is extremely important when cutting large logs.

The final cut, or release cut, will be made through the tension area. Because the offside has been cut, the sawyer only has to use enough bar to finish cutting the remaining wood. This allows the sawyer to stand back, away from the danger. The location of the pie-shaped section and the release cut vary depending on the type of bind.

Top Bind: Remove the pie-shaped section from the top, then make the release cut from the bottom.

Bottom Bind: Remove the pie-shaped section from the bottom, then make the release cut from the top.

Side Bind: If you are not certain the job is safe, do not make the cut. Normally, the offside is the side with tension; the tension side is usually bowed out (convex). Look for solid trees with no overhead hazards or other objects that you can stand behind for protection while cutting. Remove a pie-shaped section from the compression area, then make the release cut in the tension area.

End Bind: Cut from the top down, inserting a wedge as soon as possible. Finish by cutting down from the top. Watch the wood chips to make sure that the chain is not cutting in the dirt (look for dark chips).

Pay special attention when bucking in blowdown. Blowdown is a result of strong winds that have uprooted the trees. At any time while the bucking cuts are made, the roots can drop back into place or roll. Consider the following points when bucking blowdown.

Small trees growing on the roots of blowdown could be forced into the sawyer's position if the roots drop or roll. Cut the small trees off first. Limbs may be preventing the roots from rolling. Don't cut those limbs.

The roots can move in any direction. Avoid standing directly behind or downhill from them.

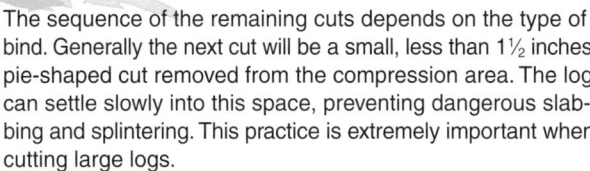

The following example shows the importance of following proper procedures when bucking blowdown.

A 30-inch d.b.h. fir tree was lying across a steep slope; the butt end was still anchored by a few roots. About 30 feet from the roots, the tree was balanced on a small stump. This stump supported the small end of the tree above some log chunks and debris.

The first step in proper bucking procedure is to inspect the log for all binds, pivot points, and skids. The sawyer failed to properly estimate the log's reaction when the log was cut. He chose downhill as his offside, expecting both the tree and roots to roll down the slope. He stood 12 feet from the roots and to the left of his saw.

When the sawyer made his release cut, the log rolled slightly uphill, off the small stump. The tree's top came to rest on the log chunks and debris, then slid rapidly downhill on these natural skids. With the stump as a pivot, the butt end swung uphill, killing the sawyer.

If the small log had been inspected thoroughly, the sawyer could have:
- Bucked the tree at or near the pivot.
- Started bucking at the small end of the tree first, leaving the roots for last.
- Stood to the right of his saw so he would be in the clear.

Safe Bucking Practices

Warn workers that are working in and below an active cutting area. Allow workers time to move to a safe location. Verify their safety visually and verbally. Announce when a bucking operation has been completed.

Buck small sections that will be easy to control when they begin moving. Removing a single section of log may require that other binds be eliminated first. Angle bucking cuts, wide on top and made on the offside, allow a single section of log to be removed. Angled cuts will permit the bucked section of log to be rolled away from the remaining log.

All logs must be completely severed when bucked. Flagging should be used to mark an incompletely bucked log as a hazard.

Never approach a cutting operation from below until the saw has stopped running, you have established communication with the sawyer, and the sawyer has granted permission to proceed.

Points to Remember

- Do a complete sizeup. Identify the hazards, and establish your escape routes and safety zones.

- Use objects such as rocks, stumps (if they are tall enough), and sound standing trees with no overhead hazards for protection in the event the tree springs sideways toward the sawyer when the release cut is made.

- Binds change with log movement. Reevaluate as necessary.

Limbing

Any of the following situations could result in a fatality or serious injury.

- Check for overhead hazards before **any** limbing begins. If a specific portion of the tree you are limbing has any overhead hazards, leave that portion of the tree unlimbed.

- Check for objects on the ground such as stumps, logs, and spring poles that may be hidden by the limbs of the felled tree. If the tip of the bar unintentionally strikes an object, the saw may kick back.

- Maintain a firm grip on the saw with your thumb wrapped around the handlebar during all limbing activities, regardless of the direction the saw is turned.

- Be sure you have firm footing as you walk down the tree that you are limbing. Calked boots are strongly recommended for walking on felled trees.

- Do not attempt to cut limbs that are supporting a tree off the ground if there's a chance the tree could roll on the sawyer. Always plan and clear an escape route.

- When limbing on top of a log, right-handed sawyers should limb the right side out to the top before turning around and limbing the other side on their way back. Another option is to limb a tree out in sections. It is not advisable for a right-handed sawyer to limb on the left side (or for a left-handed sawyer to limb on the right side). Crossing over when limbing could result in an injury.
  —The bar length should be appropriate for the sawyer's height. Proper bar length reduces bending at the waist, decreasing back strain.
  —To increase the distance between the saw chain and your leg, bend at the waist and reach out with the saw to cut the limb. This increased distance allows you more reaction time if a kickback occurs.

- Overreaching kickbacks occur when too much bar is used when cutting a limb and the bar tip contacts an object, propelling the bar up and back (figure 3-7).

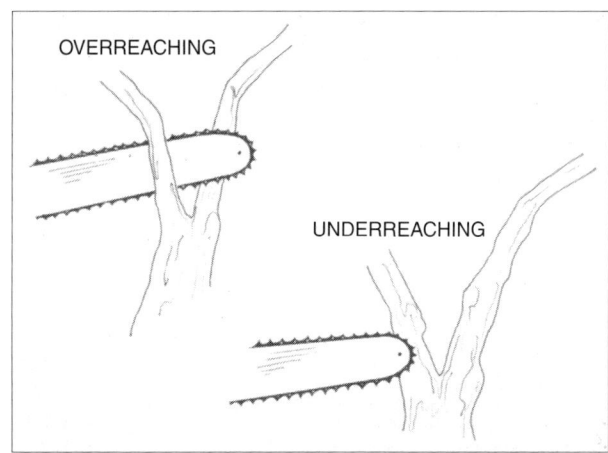

Figure 3-7—Avoid overreaching and underreaching kickbacks by using the bar properly.

- Underreaching kickbacks occur when too little of the bar tip is used and the tip contacts any object, propelling the bar back and down.

- Spring poles are limbs or small trees that are bent over and under extreme tension. Spring poles are encountered frequently when limbing. They can cause serious injury. If spring poles are not cut properly, they can spring back and strike the sawyer or throw the chain saw back into the sawyer. Sawyers must recognize spring poles and use the proper technique when cutting them. Stand back at a safe distance and make a series of shallow cuts on the portion of the spring pole that is being compressed. Make your release cut in the top side of the portion of the spring pole that will fall to the ground.

- Sometimes a tree is suspended off the ground by the limbs underneath or by uneven terrain. The sawyer must decide whether or not to limb the tree after considering the potential that the sawyer might fall or that the tree may roll or collapse. Consider footwear and environmental conditions such as rain, snow, fog, or darkness, and the ability and experience of the sawyer. Carefully select the appropriate technique, such as limbing from the ground, limbing on top, or lowering the tree by bucking (figure 3-8).

Figure 3-8—The top side of suspended logs should be limbed on the top side while you are standing on the ground or lower the log to the ground by bucking.

Brushing and Slashing

Sizeup and Safety Considerations

Many sawyers have cut their chaps or their legs when they took a step toward the next tree. Be sure the chain has stopped before moving to the next tree. Engage the chain brake when moving short distances.

Shut the saw off when moving farther than from tree to tree, when hazardous conditions exist (slippery surfaces or heavy underbrush), and whenever moving more than 50 feet.

When slashing (felling) trees smaller than 5 inches in diameter, an undercut may not be needed. Instead, a single horizontal cut (kerf face) one-third the diameter of the tree may be used to fell it. When directional felling is necessary, use a normal face cut (consisting of a horizontal and sloping cut or two sloping cuts). Situations when directional felling should be used include:

- A potential barber-chair situation.
- A closed canopy.
- Tree defects.
- Side binds.
- Environmental damage.

Other Mitigating Circumstances—Always escape the stump quickly even when felling small-diameter trees. They can cause serious injuries and fatalities.

Trees should be pushed over only by the sawyer, only when the sawyer can do so safely, and only after the sawyer has looked up for overhead debris that could become dislodged.

Safe and Efficient Brushing and Slashing Techniques

In dense fuel accumulations, the tip of the guide bar may accidentally bump (stub) into a limb. The sawyer must continually be alert for kickback.

The sawyer normally will have a hand piler working nearby helping to remove cut debris. The piler's safety must be taken into consideration.

Proper stance and saw handling is imperative. In addition, the following steps should be taken.

- *LOOK UP* for widow makers and other loose debris. Don't cut under a hazard. Remove the hazard or move the cutting location.

- If possible, stay on top of logs while limbing. Doing so reduces the chance that the log will roll over on the sawyer if the sawyer cuts a supporting limb.

- Watch out for whipping limbs and branches when cutting smaller material. Cut close to the stem. Begin and complete cuts with a sharp chain and high chain speed. Use eye protection.

- Cut limbs and stems flush with the trunk or close to the ground. Do not leave pointed stems that could cause injury during a fall or cause the sawyer to trip.

- Don't cross the chain saw in front of your legs. Keep some distance between your legs and the guide bar. Bend down to maintain distance. Cut on one side, then the other to avoid crossing the chain saw in front of you.

- Never cut with the chain saw above shoulder height (figure 3-9). Control is difficult when the saw's weight is above your shoulders. A thrown chain could strike you in the face or upper body.

Chapter 3—Chain Saw Tasks and Techniques

Figure 3-9—Do not cut with the chain saw above shoulder height. Keep the chain saw below your shoulders to maintain control when cutting.

discussed the plan, work systematically from the outside in and from downhill up. This reduces the chance that material will hang up. Maintain a space between workers that is at least two-and-one-half times the height of the tallest tree.

- You need one or more escape routes, even when felling small trees.

- Begin to develop a felling and bucking pattern as you work into the area. As you down more material, be increasingly careful of your footing and continually identify new escape routes.

- Special hazards like leaners and snags need to be taken care of right away. Get hazard trees on the ground so no one has to work under them (figure 3-10).

Figure 3-10—Hazard trees, such as leaners, need to be removed to prevent anyone from working under them.

- Clear debris from the cutting location to prevent the guide bar tip from stubbing it accidentally. When you are removing debris, engage the chain brake or turn off the ignition.

- Watch out for spring poles. Do not cut spring poles if you can avoid doing so. If you must cut a spring pole, make your release cut in the top side of the portion of the spring pole that will fall to the ground. Be careful not to stand in the path of the pole when tension is released.

- When cutting a heavy limb, consider using a small cut opposite the final cut to prevent the material from slabbing or peeling off.

- Pay special attention if you are working in close quarters with other workers in an area with steep slopes and thick brush or logging slash. First, stop and size up the situation. Make a plan and talk it over with all workers in the area. A well thought-out plan saves time and reduces the risk of accidents. After you've

- Double stumps (figure 3-11) are hazardous because they have a high potential for causing kickback. Watch that bar tip!

- When felling small trees, cut the stumps as close to the ground as possible without hitting the dirt with your chain. Stumps are cut low so they will not be as noticeable and will present fewer hazards for people and wildlife.

Chapter 3—Chain Saw Tasks and Techniques

Basic Felling

Situational Awareness

Analyze the felling job by considering:
- Species (live or dead).
- Size and length.
- Soundness or defects.
- Twin tops.
- Widow makers and or hangups.
- Frozen wood.
- Rusty (discolored) knots.
- Punky (swollen and sunken) knots.
- Frozen wood.
- Footing.
- Damage by lightning or fire.
- Spike top.
- Heavy snow loading.
- Bark soundness.
- Direction of lean.
- Degree of lean (slight or great).
- Head lean or side lean.
- Nesting or feeding holes.
- Splits and frost cracks.
- Deformities, such as those caused by mistletoe.
- Heavy branches or uneven weight distribution.

Analyze the base of the tree for:
- Thud (hollow) sound when struck.
- Conks and mushrooms.
- Rot and cankers.
- Shelf fungi or "bracket".
- Wounds or scars.
- Split trunk.
- Insect activity.
- Feeding holes.
- Bark soundness.
- Resin flow on bark.
- Unstable root system or root protrusions.

Examine surrounding terrain for:
- Steepness.
- Irregularities in the ground.
- Draws and ridges.
- Rocks.
- Stumps.
- Loose logs.
- Ground debris that can fly or kick up at the sawyer.

Examine immediate work area for:
- People, roads, or vehicles.
- Powerlines.
- Hang ups and widow makers.
- Other trees that may be affected.
- Fire-weakened trees.
- Hazards such as trees, rocks, brush, low-hanging limbs.
- Reserve trees.
- Structures.
- Openings to fall trees.
- Snags.
- Other trees that may have to be felled first.

Figure 3-11—Double stumps have a high potential to cause kickback.

- Small trees can be limbed while they are standing. Don't cut with the chain saw above shoulder height. Limbing the bottom of small trees allows the sawyer to move in closer to the bole when felling it and will help the sawyer watch the tip of the bar to prevent kickbacks.

- The chain is more likely to be thrown when you are working with small material. Check your chain tension often. Sawing close to the ground increases the chances of kickback and damage to the chain. Watch out for rocks and other debris.

- Remember, when you saw up from the bottom (using the top of the bar) the saw will push back rather than pull away. This increases the risk of kickback and loss of control. Be aware of signs of fatigue like more frequent kickbacks, bar pinches, and near misses. Take a break when you show signs of fatigue.

- Cut pieces small enough so they are easy to lift and handle. Lift properly using the legs and keeping the back straight. Hand pilers must be aware of their footing and watch out for flying debris.

Walk out and thoroughly check the intended lay or bed where the tree is supposed to fall. Look for dead treetops, snags, and widow makers that may cause kickbacks, allow the tree to roll, or result in another tree or limb becoming a hazard. The escape route and alternate routes must be predetermined paths where the sawyer can escape once the tree is committed to fall or has been bucked. Safe zones should be no less than 20 feet from the stump, preferably behind another tree that is sound and large enough to provide protection. Escape routes and safe zones should be 90 to 135 degrees from the direction of fall. Sawyers must select and prepare the work area and clear escape routes and alternate routes before starting the first cut.

Sizeup

Before starting the saw, the chain saw operator must be able to evaluate if a tree is safe to cut. Other options are always available. *IF FELLING A PARTICULAR TREE IS DANGEROUS, DON'T DO IT!*

Debris falling from above causes most accidents. Practice watching overhead while cutting, with occasional glances at the saw, the kerf, and the top of the tree.

Observe the Top—When you approach the tree to be felled, observe the top. Check for all overhead hazards that may come down during felling.

Look at the limbs. Are they heavy enough on one side to affect the desired felling direction? Do the limbs have heavy accumulations of ice and snow?

Are the limbs entangled with the limbs of other trees? If so, they will snap off or prevent the tree from falling after it has been cut.

Is the wind blowing strong enough to affect the tree's fall? Wind speeds higher than 15 miles per hour may require that felling be stopped. Strong winds could also blow over other trees and snags in the area. Switching or erratic winds require special safety considerations.

Check For Snags—Check all snags in the immediate area for soundness. A gust of wind may cause snags to fall at any time, as may the vibration of a tree fall. If it is safe to do so, begin by falling any snag in the cutting area that poses a threat.

Swamp Out the Base—Clear small trees, brush, and debris from the base of the tree. Remove all material that could cause you to trip or lose your balance. Also remove material that will interfere with the saw, wedges, and ax. Look for small trees and brush that could accidentally stub the guide bar. Be careful not to fatigue yourself with unnecessary swamping. Remove only what is needed to work safely around the base of the tree.

Assess the Tree's Lean (figure 3-12) and the Soundness of the Holding Wood—Most trees have two natural leans; the predominant head lean and the secondary side lean. The leaning weight of the tree will be a combination of these two leans. Both must be considered when determining the desired felling direction. The desired felling direction can usually be chosen within 45 degrees of the combined lean, provided there is enough sound holding (hinge) wood to work with, especially in the corners of the undercut. Evaluate the tree's lean. With a plumb bob or ax, project a vertical line up from the center of the tree's butt and determine if the tree's top lies to the right or left of the projected line.

A pistol-grip tree may appear to be leaning in one direction while most of the weight is actually in another direction (figure 3-13).

Look at the tree top from at least two different spots at right angles to each other. This will be done again in the sizeup process.

The importance of the holding wood cannot be overemphasized. Determine the condition of the holding wood by sounding it with an ax. Look up for falling debris while doing so.

Boring is an important technique, but it must be done properly because it has the potential for kickback. Using the guide bar tip, bore vertically into the area immediately in front of and behind the holding wood (figure 3-14). Do not weaken the holding wood by boring into or across any of the holding wood. The color of the sawdust and ease with which the saw enters the wood will be your indication of the tree's soundness. Begin the boring cut with the power head lower than the tip of the bar. After the tip is in the tree deep enough to prevent it from kicking back, apply full throttle. The thumb should be wrapped securely around the handlebar. Maintain full throttle throughout the boring cut.

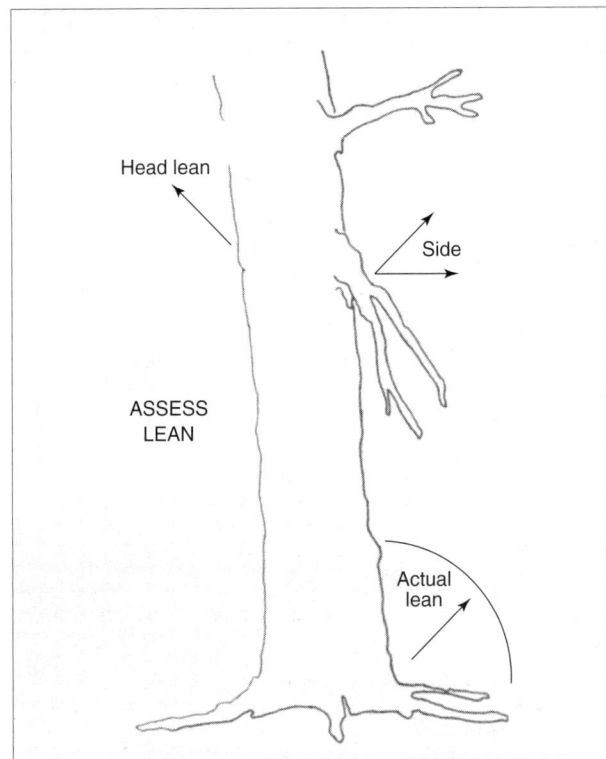

Figure 3-12—Assess the tree for head lean and side lean.

Check for frost cracks or other weak areas in the holding wood. The desired felling direction can be adjusted to eliminate weaknesses in the holding wood. The depth of the undercut can also be adjusted (less than one-third the tree's diameter or greater than one-third) so that the holding area takes advantage of the soundest wood available.

Escape Routes

With the desired felling direction in mind, determine your escape route (figure 3-15). Consider which side of the tree you will be making your final cut on and select a path that will take you at least 20 feet behind the stump when the tree begins to fall. Don't choose a path directly behind the tree. It is best to prepare two escape routes in case you switch your location on the final cut.

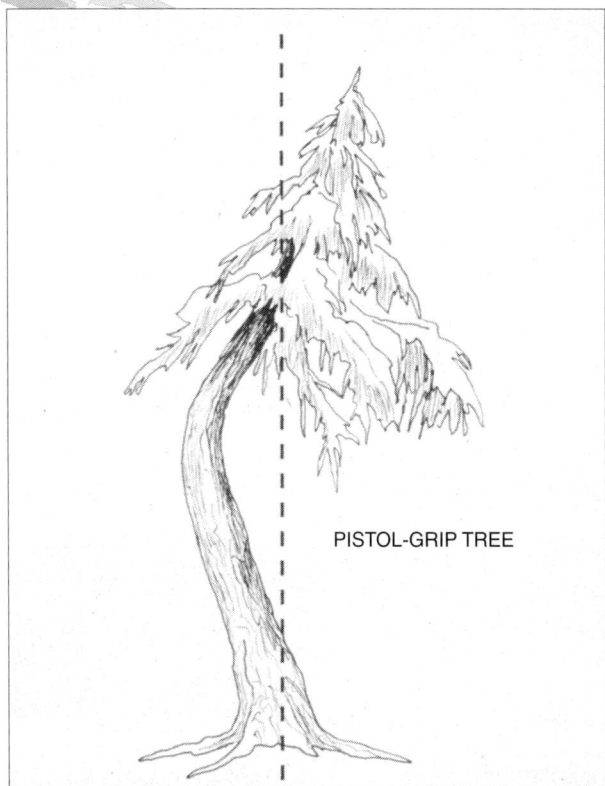

Figure 3-13—The lean of a pistol-grip tree is hard to determine.

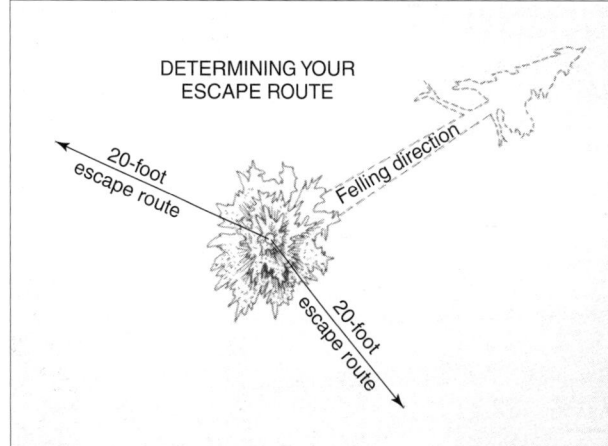

Figure 3-15—Keep the felling direction in mind when planning escape routes.

Look for a large solid tree or rock for protection. The tree or rock must be at least 20 feet away from the stump and not be directly behind it. Make sure that debris that could trip you is cleared from the escape route. Practice the escape.

Walk out the intended lay of the tree (figure 3-16). Look for any obstacles that could cause the tree to kick back over the stump or cause the butt to jump or pivot as the tree hits the ground. Look for any small trees or snags that could be thrown into your escape route. Check to be sure the cutting area is clear of people.

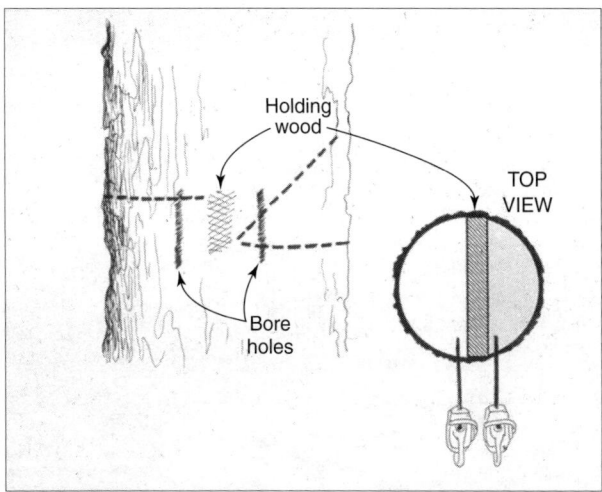

Figure 3-14—Do not weaken the holding wood by boring into or across any of the holding wood.

Figure 3-16—Check the intended lay of the tree for unwanted obstacles.

Using the observations you made walking out the lay, reexamine the escape route. Be sure that your chosen route will be the safest escape—before you begin to cut.

Felling the Tree

"Face" the tree (figure 3-17). The face is made in the direction you want the tree to fall. Estimate one-third of the tree's diameter, shout a warning if necessary, and proceed with the undercut. When finished, check the direction the tree is faced. If the face is not in the desired felling direction, correct the cut.

Figure 3-17—Check the direction the tree is faced.

Shout a warning (figure 3-18). Shut off the saw and shout to be sure that the cutting area is secure. Reexamine your primary and secondary escape routes before beginning the backcut.

Figure 3-18—Shout a warning to make sure the cutting area is secure.

Complete the backcut (figure 3-19). Remembering the importance of holding wood, stump shot (see glossary), and wedging, complete the backcut.

Escape the stump. When the tree commits to the fall, rapidly follow your escape route. Do not hesitate at the stump. If your saw becomes stuck, leave it. If carrying the saw prevents you from escaping quickly enough, drop it!

Keep your eyes on your predetermined escape route. If the felled tree strikes other trees, they may still be moving after the tree has fallen. Watch for flying limbs and tops. Remain in your safety zone until it is safe to approach the stump.

Figure 3-19—Complete the backcut.

Analyze the operation (figure 3-20). The stump gives the best critique of the felling operation. Before approaching the stump, look in the tops of the surrounding trees for new overhead hazards.

Take a moment or two to look at the stump. Does the tree have the desired lay? How much holding wood is left on each corner? Is the stump shot sufficient? Were the cuts level? Take time to analyze the felling operation. Check stump height and look for stump or root pull and dutchman cuts.

Figure 3-20—Look at the stump when critiquing a felling operation.

Felling Details

Proper evaluation of a felling operation requires a thorough understanding of the mechanics of the undercut, holding wood, backcut, and the felling procedure. In addition, the sawyer needs to consider the various tree problems. Every tree should be evaluated or sized up using techniques discussed in *Situational Awareness—Evaluating the Complexity of the Assignment* in chapter 1.

The Undercut, Holding Wood, and Backcut

The three basic cuts are the conventional undercut, the Humboldt, and the open face. We are only going to discuss the conventional undercut because of its broad application for all timber types and because it provides a solid foundation from which to learn additional cutting techniques.

It takes three cuts to fell a tree. Two cuts form the undercut (or face cut) and the third forms the backcut (figure 3-21). The correct relationship of these cuts results in safe and effective tree felling. Before discussing the felling procedure we will analyze the mechanics of the felling cuts. Undercutting and

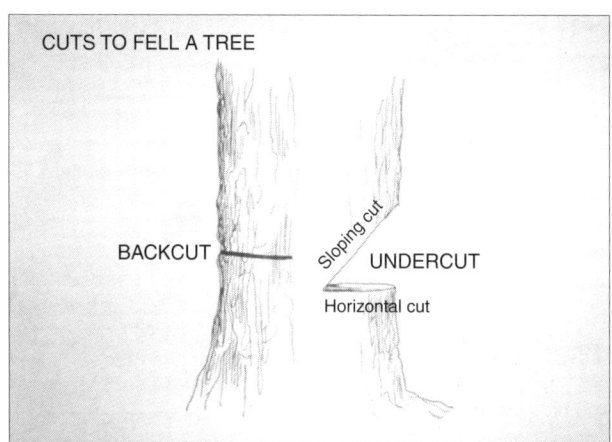

Figure 3-21—An undercut and a backcut are required to fell a tree. A horizontal cut and a sloping cut make up the undercut. The backcut is the third cut needed to fell a tree.

backcutting construct the hinge that controls the direction and fall of the tree.

The undercut serves three purposes. First, it allows the tree to fall in the chosen direction by removing the tree's support in the direction of the face. Second, it enables control because the tree slips off the stump, rather than jumping off. Third, when the tree is breaking the holding wood, the undercut prevents the tree from kicking back.

The undercut consists of two cuts, a horizontal cut and a sloping cut. Observe overhead hazards and look up often during the undercut. You should either be down on one knee or standing all the way up.

The tree is faced in the general direction of the tree's lean. Ideally, the undercut is made in the same direction as the tree's lean, but because of structures, roads, other trees, or trails, the desired felling direction may be to one side or the other of the lean. Normally, the desired direction is less than 45 degrees from the lean.

The horizontal cut is a level cut. This cut is made close to the ground unless a snag is being felled or another factor creates special hazards for the sawyer. The horizontal cut dictates the direction of fall if the relationships of the three cuts are maintained. If there is any danger from above, such as snags, the cutting should be done while standing so the sawyer can watch the top and escape more quickly. After selecting the desired felling direction, estimate one-third the tree's diameter, set the saw's bumper spikes at this point, and begin the horizontal cut.

The specific direction of the undercut is determined by "gunning" the saw. Look down the gunning marks on the saw and align them with the desired felling direction. After the cut has been made level to at least one-third of the tree's diameter, the horizontal cut is complete. Short snags sometimes require an undercut deeper than one-third the tree's diameter to offset the tree's balance. Felling short snags will be addressed later. Trees with heavy leans may not allow the sawyer to make the horizontal cut as deep as one-third of the tree's diameter without pinching the guide bar.

When the horizontal cut is complete, remove the bark from an area on both sides of the kerf. The bark can be removed with your ax or with the tip of the guide bar (figure 3-22). Watch out in case the ax glances off the bole or the saw kicks back.

Figure 3-23—The sloping cut is a 45-degree angle.

Figure 3-22—Bark can be removed with the tip of the bar.

Figure 3-24—When the sloping cut and the horizontal cut cross, a dutchman is formed.

The sloping cut needs to be angled so that when the face closes the tree is fully committed to the planned direction of fall. As the face closes, the holding wood breaks. If this happens and the tree is still standing straight, the tree could fall away from the desired direction.

As a general rule, make the sloping cut at a 45-degree angle (figure 3-23). Remember that it is important that the face not close until the tree is fully committed to the planned direction of fall.

Line up the sloping cut with the horizontal cut so that they meet, but do not cross. When the cuts cross, a "dutchman" (figure 3-24) is formed. If the tree were felled with a dutchman, first the dutchman would close, then the tree would split vertically (barberchair), or the holding wood would break off. Felling control would be lost. A weak tree might snap off somewhere along the bole or at the top. It is difficult to make the sloping cut and the horizontal cut meet correctly on the opposite side of the tree. This is because the point of intersection is not immediately visible to the sawyer.

After making a short sloping cut, leave the saw in the cut. Go around to the other side of the tree and see if the guide bar is in the correct plane to intersect the back of the horizontal cut. Keep your hands away from the throttle trigger. Engage the chain brake.

Practicing on high stumps will help you become skilled at lining up these cuts.

The holding wood is the wood immediately behind the undercut. The most important portion of the holding wood is in the very corners of the cut, in the first 4 to 8 inches inside the bark. The horizontal and sloping cuts must not overlap in this region. If they do, the undercut must be cleaned up so no dutchman is left in these corners. Care must be taken not to cut the undercut too deeply while cleaning up. This reduces the amount of room available for wedges.

If cleaning up the sloping cut will create too deep an undercut, stop the sloping cut directly above the end of the horizontal cut.

The undercut needs to be cleaned out. Any remaining wood will cause the face to close prematurely and the holding wood will be broken behind the closure.

Once the face has been cleaned, recheck the felling direction. Place the saw in the face and check the gunning marks (figure 3-25) or stick an ax head into the face and look down the handle. The back of the undercut should be perpendicular to the desired felling direction.

Figure 3-25—Use the gunning marks to check the felling direction.

If the tree is not aimed in the direction that you want it to fall, extend the horizontal and sloping cuts as needed, maintaining a single plane for each of the two cuts.

Backcut and Wedging Procedures—The third cut needed to fell a tree is the backcut. The relationship of this cut to the face is important for proper tree positioning and the sawyer's safety. The backcut can be made from either side of the tree if the saw has a full-wrap handlebar, as recommended in the *Health and Safety Code Handbook*. Choose the safest side to cut on (figure 3-21).

In the area where you have removed the bark behind the horizontal cut, place the bumper spikes so the chain will cut no closer than 2 inches from the face and 2 to 5 inches, depending on tree size, above the face's horizontal cut. You may place the bumper spikes closer than 2 inches on smaller trees with lightweight tops.

The best way to envision these cuts is by the use of a rectangle (figure 3-26). The bottom corner is the back of the face's horizontal cut. The opposite upper corner will be the back of the backcut.

The height of the rectangle is referred to as the stump shot. It is an antikickback device to prevent the tree from kicking back over the stump if it hits another tree during its fall. This is especially important felling trees through standing timber.

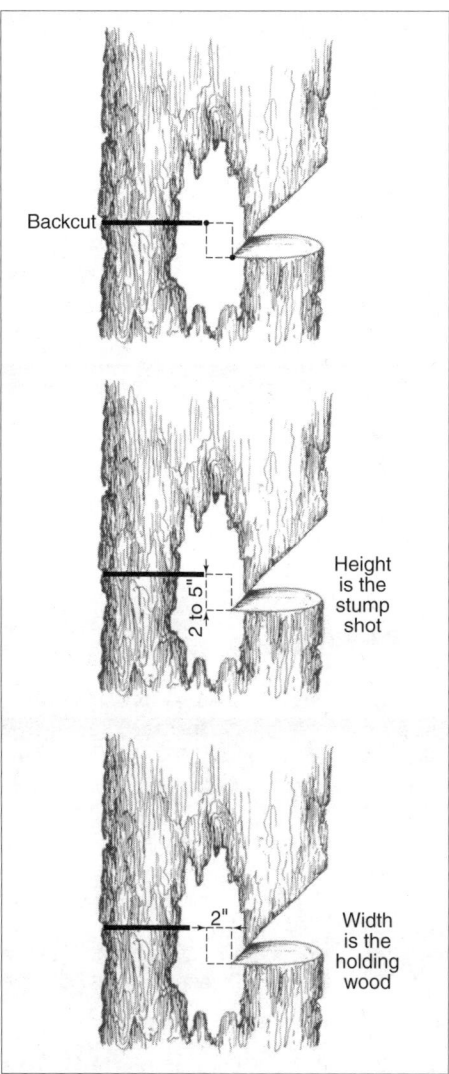

Figure 3-26—An imaginary rectangle can help the sawyer understand the importance of the backcut.

The width of the rectangle is the holding wood. As the backcut is made, the sawyer must be careful not to cut this wood. Maintaining the holding wood is the key to safe and effective felling.

Start the backcut with the bumper spikes placed so the chain will end in the upper corner of the rectangle. Hold the saw level so that the backcut will be level when the cut is complete. You want to be sure that when the cut is finished it will line up with the top corner of the opposite rectangle. If the cut is angled, wedging power and/or the stump shot's height will be altered.

It may be helpful to cut or chop the bark to help level the bar. Do not cut deeper than the bark. Cuts into the wood will eliminate or reduce wedging lift. Once the cut has been made into the wood, do not change the cut's location.

Keep at least three wedges and an ax readily accessible while making the backcut. The wedges should be in a wedge holster worn on the waist or in pants pockets. Keep the ax within arm's reach. The size of the wedge depends on the tree's diameter. For a 24-inch tree, two 10- to 12-inch wedges and one 4- to 6-inch wedge is a good combination.

If there is any wind at all, at least two wedges are recommended. The second wedge adds stability. With only one wedge, the tree can set up a rocking action between the holding wood and the wedge. If a strong wind begins to blow, the holding wood can be torn out.

Remove thick bark immediately above and below the backcut's kerf where wedges will be placed. The bark will compress, lessening the lifting power of the wedges. The wedges should be spread to better stabilize the tree in case of erratic winds.

Directional Felling

Place wedges in the kerf in each area as soon as the chain will permit. Place the wedges parallel to the desired felling direction. Do not drive the wedges too hard. They will interfere with the backcut or cause the tree to become a heavy leaner.

Watch for droop in the wedges and occasionally try to push them in with your hands. Retighten them with an ax every 4 to 6 inches of cutting. Be careful not to drive the wedges too hard. The wedges are there in case the lean was incorrectly established, the wind causes the tree to set back, or the sawyer intends to fell the tree in a different direction from the tree's natural lean.

As you cut, continually look above for possible hazards and at the kerf for movement. Do not cut the holding wood.

The gunning sights can be used in reverse to help determine the guide bar's position. With the correct lean established and the proper relationship between the three cuts, the face will begin to close and the tree will fall in the planned direction.

Wedges must be used for all felling operations. Small trees limit the sawyer's use of wedges, even when small wedges are used. A technique can be employed where half of the back cut is made at a time. This allows for wedges to be placed without interfering with the guide bar.

After making the undercut, cut half of the backcut using the guide bar's tip. Make this cut from the tree's offside. Watch out for kickback and be careful not to cut the holding wood. Finish the backcut from the other side. Leave 1 to 1½ inches of holding wood.

After removing the saw, place a small wedge in the kerf an inch or more from the remaining wood to be cut. Remember to keep the wedge tight but do not drive the wedge too hard. Finish the backcut using the tip of the guide bar, being prepared for kickback from the wedge. The wedge will be in position if the tree sits back. If two wedges are needed to lift the tree into the undercut, spread them as wide as possible (figure 3-27). For trees that have a moderate amount of side lean, two wedges may be inserted on the side of the backcut that has been cut first. As with other wedging operations, it is essential to tighten the wedges often, especially on trees that are attempting to sit back.

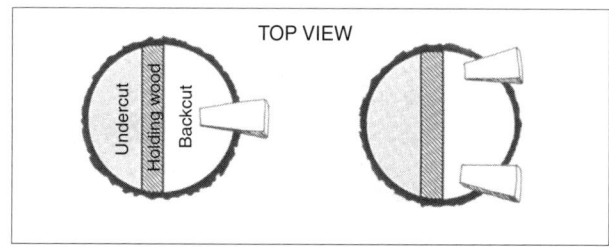

Figure 3-27—Wedges need to be parallel to the desired felling direction.

A sitback is a tree that settles back opposite the intended direction of fall during the backcut. This normally happens because the lean was incorrectly established or the wind changed. If the sawyer has been following the proper felling procedure (there is a wedge in the backcut and the holding wood has been maintained), a sitback can be dealt with readily. If the proper felling procedure has not been followed, the sawyer will need assistance because the tree has probably pinched the bar (figure 3-28).

Chapter 3—Chain Saw Tasks and Techniques

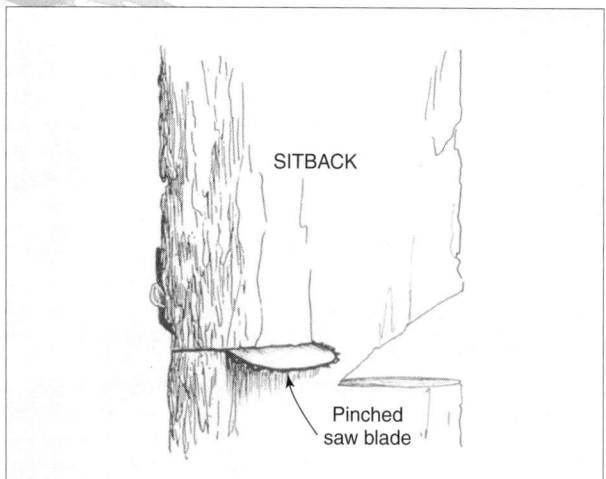

Figure 3-28—A tree settles back opposite the intended direction of fall.

Immediately notify your supervisor in the case of sitbacks and other felling difficulties. Before you leave a hazardous tree, be sure to clearly mark the area with flagging (figure 3-29) or with a written warning. Give a verbal warning to others working in the area.

Figure 3-29—Clearly mark a hazard tree before leaving the area. Notify your supervisor of the hazard.

A snag is a standing dead tree, or portion of a tree, whose wood is decomposing.

Shout a warning. Before working on a snag, everyone in the area must be notified. Remember a snag can fall in any direction at any time.

Observe the top. Pay special attention to overhead hazards, branches, and the snag's top. Upper limbs may be weak and ready to come down at the least vibration. Never cut directly below a hazard. Look up while driving wedges.

Swamp out the base. Carefully check the condition of the bark on the snag. Loose bark can come sliding down the side of the snag and presents an extreme hazard to the sawyer. Standing back with room to escape, remove loose bark at the snag's base by prying it with an ax or a pole. Do not chop the bark, because this would set up vibration in the snag.

Size up. Check the condition of wood by boring into it with the bar tip. Maintain the integrity of the holding wood. When sounding with an ax, look up while striking the tree. Check for frost cracks and other splits in the holding wood.

Determine two escape routes. Since the holding wood is rotten to some degree, you must establish two routes of escape.

Make the undercut. Do not fell a snag against its lean. Make the undercut and the backcut while standing upright. You are in a position where you can easily look up. In addition, you are in a position that allows immediate escape.

When you are cutting the face, be alert for the snag pinching the bar. Previous boring in the undercut area during sizeup should alert you to this possibility. Moving the bar back and forth will minimize the possibility of pinching. If the snag starts to sit on the bar, finish the undercut just to that depth. It is critical that the undercut has a wide opening and that it be cleaned out from corner to corner.

A short snag, with few or no limbs to give it lean, may need a face up to one-half the snag's diameter to offset the balance.

Felling Observers and Spotters

The use of personnel other than swampers in felling operations has been controversial. The *Health and Safety Code Handbook* states that if you choose to have additional personnel (such as during training), justification for the additional personnel and the implementation process shall be documented in the JHA.

Chapter 4—Crosscut Saw Tasks and Techniques (Suggested time: 2 hours)

Understanding Your Crosscut Saw

After completing this section, students will:

- Recognize a quality vintage saw and understand different crosscut features.
- Describe how a saw cuts and how the saw's components function.
- Test a crosscut saw, assess its performance, and recommend maintenance measures to correct any deficiencies.
- Demonstrate knowledge of saw handles and their relationship to saw performance in various applications.
- Develop a good saw maintenance program and understand how to maintain a good working relationship with the saw filer.
- Demonstrate an understanding of saw sheathing and transport requirements.

Historical Origin of the Crosscut Saw

The crosscut saw did not come into use until the 15th century. Early saws had a plain peg-tooth design. Saws were used in Colonial America and were being manufactured in this country by the mid-1800's. Saws were not used for felling timber until around 1880. The machinery to make these vintage saws is no longer available. Crosscut saws manufactured today are lower quality.

Different Types of Crosscut Saws

Crosscut saws can generally be divided into two types: one- and two-person saws.

One-Person Crosscut Saws

A one-person crosscut saw's blade is asymmetrical. The saw has a D-shaped handle. The saw also has holes for a supplemental handle at the point (tip) and the butt (near the handle). The saws are usually 3 to 4½ feet long (figure 4-1).

Figure 4-1—One-person crosscut saw.

Two-Person Crosscut Saws

Two-person crosscut saws are symmetrical. They cut in either direction on the pull stroke. Vintage saws were made from 4 to 7 feet long in 6-inch increments.

Two-person crosscut saws (figure 4-2) manufactured today are flat ground. Most vintage saws were either straight taper, crescent taper, or flat ground. The saws have one or two holes, or a groove, on the blade ends to attach removable handles. Most vintage saws had teeth all the way to the ends of the blade. Saws manufactured today do not.

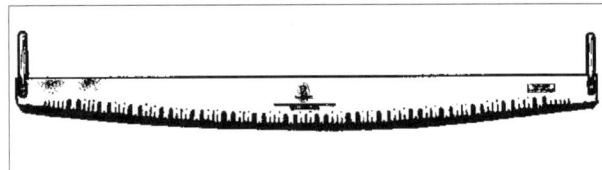

Figure 4-2—Two-person crosscut saw.

Two-Person Crosscut Saw Patterns

Felling Saws

Felling saws (figure 4-3) are best suited for working in a horizontal position. Felling saws have a concave back and are narrower than bucking saws. The combination of a concave back and narrower width give felling saws the following characteristics:

- The saw is more flexible.
- The saw is lighter, so less effort is needed to use it.
- The sawyer can insert a wedge sooner.

Many vintage felling saws have only a single handle hole in each end.

Bucking Saws

Bucking saws (figure 4-3) can be used for felling. Some saws were manufactured to try to incorporate the best characteristics of both types of saws. Bucking saws have a straight back; they are much thicker than felling saws, so they are heavier and stiffer.

Because the bucking saw is usually operated by one person, it cuts on both the push and pull strokes. The saw's additional stiffness helps prevent the saw from buckling on the push stroke.

Chapter 4—Crosscut Saw Tasks and Techniques

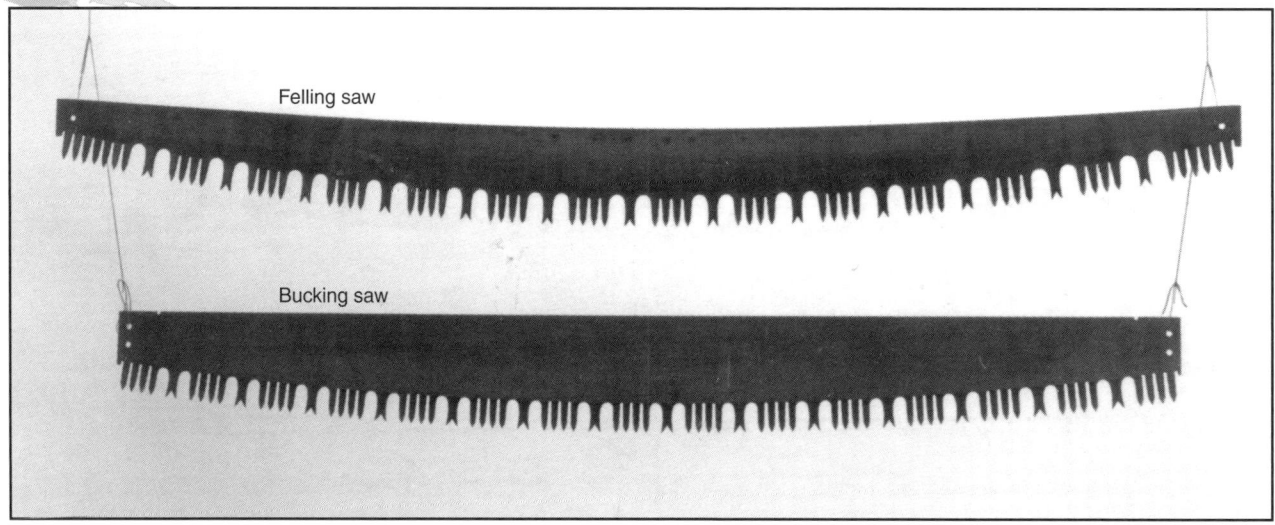

Figure 4-3—Comparison of felling and bucking saws.

Because felling saws are flexible, they do not make a good bucking saw or a general all-around utility saw. The bucking saw is recommended as the standard saw for most trail and construction applications today.

Saw Grinds

Historically, the sides of a saw were ground using one of three methods. Each method affected the thickness of the saw. These methods are flat, straight taper, and crescent taper.

Flat Ground

On a flat-ground saw, the metal's thickness is the same throughout. Saws manufactured today are flat ground.

Straight Taper Ground

Straight taper-ground saws have an advantage over a flat-ground saw because the saw is thinner at the back than at the center (figure 4-4). The back of the saw has more clearance, reducing binding.

Straight taper-ground saws require less set. Set is the cutter tooth's offset from the plane of the saw.

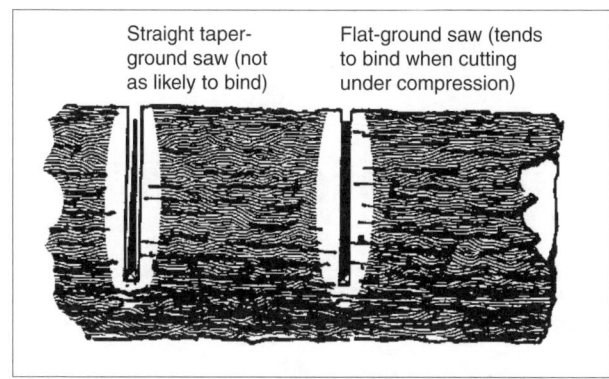

Figure 4-4—Straight taper-ground and flat-ground cuts in a log.

Crescent Taper Ground

The best vintage saws were crescent taper ground (figure 4-5). Early saw manufacturing companies used different trade names for crescent taper-ground saws. The names included: crescent ground (Simonds), improved ground (Disston), and segment ground (Atkins).

Crescent taper-ground saws offer the saw the most clearance in the kerf of any of the grinds. These saws require the least amount of set, allowing the narrowest kerf. The thinnest part of a crescent taper-ground saw is at the back center.

Chapter 4—Crosscut Saw Tasks and Techniques

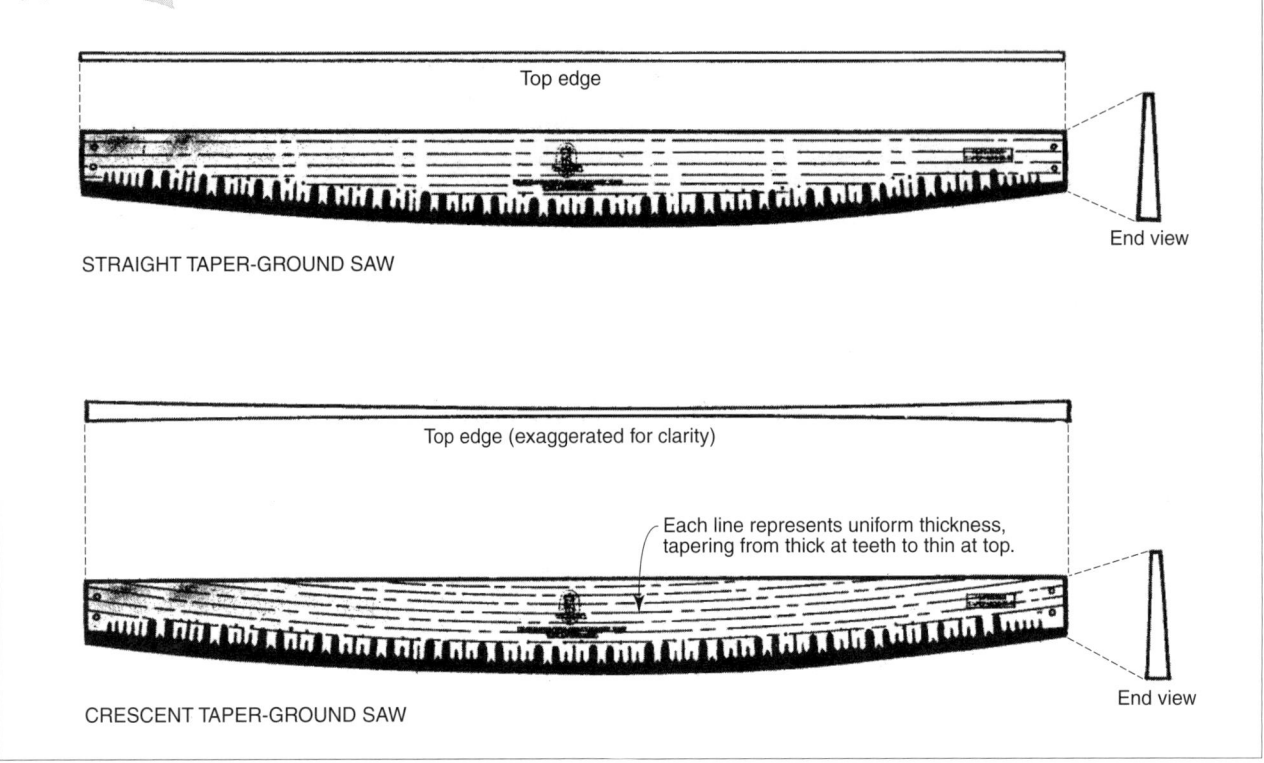

Figure 4-5—Comparison of straight and crescent taper-ground saws.

How a Saw Cuts

A saw functions like a series of knives (teeth) making simultaneous parallel cuts and releasing the wood between them.

Cutter Teeth

All saws, regardless of the tooth pattern, are made up of two rows of cutting edges. The saw releases wood fibers on each side of the kerf as it passes through a log (figure 4-6).

Cutters work best in brittle, seasoned wood. The weakened fiber is easily removed.

Rakers

Wet or green wood is hard to remove from the kerf because it is resilient. Even when the fiber is dislodged, it clogs a saw's cutter teeth.

A special kind of tooth, the raker, allows the cutter teeth to work more effectively with less effort. Even though the rakers do not sever fiber, they do perform the other two functions of saw teeth: breaking loose the cut fiber and removing it from the log. Rakers remove material whether the saw is being pushed or pulled.

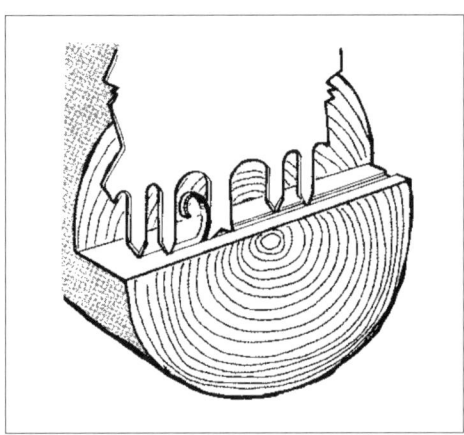

Figure 4-6—How a saw cuts.

Chapter 4—Crosscut Saw Tasks and Techniques

Gullets

Wood fiber that has been severed must be stored by the saw while it is moved through the kerf and out of the cut. This storage area (the largest space between cutters or groups of cutters) is called a gullet.

The gullet must be large enough to store all the shavings until the gullet clears the log and the shavings fall free.

The gullets (figure 4-7) determine the length of saw to use for a given application. Example: A gullet in the middle of a 3-foot log must travel 1½ feet to clear its shavings on either side. At least a 6½- or 7-foot saw would be needed to provide this travel.

Tooth Patterns

For centuries, only the plain-tooth (or peg-tooth) pattern was used. Modifications to the plain-tooth pattern were developed to make the work easier. We will discuss six patterns: the plain tooth, the M tooth, the great American tooth, the champion tooth, the perforated lance tooth, and the lance tooth (figure 4-8).

Plain-Tooth (Peg-Tooth) Pattern

This pattern just includes cutter teeth. It is best used for cutting dry, very hard, or brittle small-diameter wood. Examples include many bow saws and pruning saws.

M-Tooth Pattern

This is the second generation of saw tooth patterns. The tooth pattern consists of pairs of teeth separated by a gullet. The outer edges of the teeth (the legs of the M) are vertical and act like rakers. The inside edges of the M are filed to a bevel, making a point.

Great American-Tooth Pattern

This tooth pattern, three teeth separated by a gullet, is designed to cut dry, medium-to-hard woods. A special file is used to file these saws. The file can still be purchased today and is called a crosscut file or a Great American file.

Champion-Tooth Pattern

This pattern is especially popular in the hardwood regions of North America. It consists of two alternately set cutter teeth and an unset raker with a gullet between them. The cutters are wider and more massive than the lance-tooth pattern, allowing heavy sawing in extra hard, dry, or frozen wood. The larger teeth are sharpened in more of an almond shape rather than the pointed shape of a lance tooth.

Perforated Lance-Tooth Pattern

This tooth pattern is considered a general utility pattern that can cut all but hard and frozen wood. It consists of groups of four alternately set cutters separated by an unset raker with

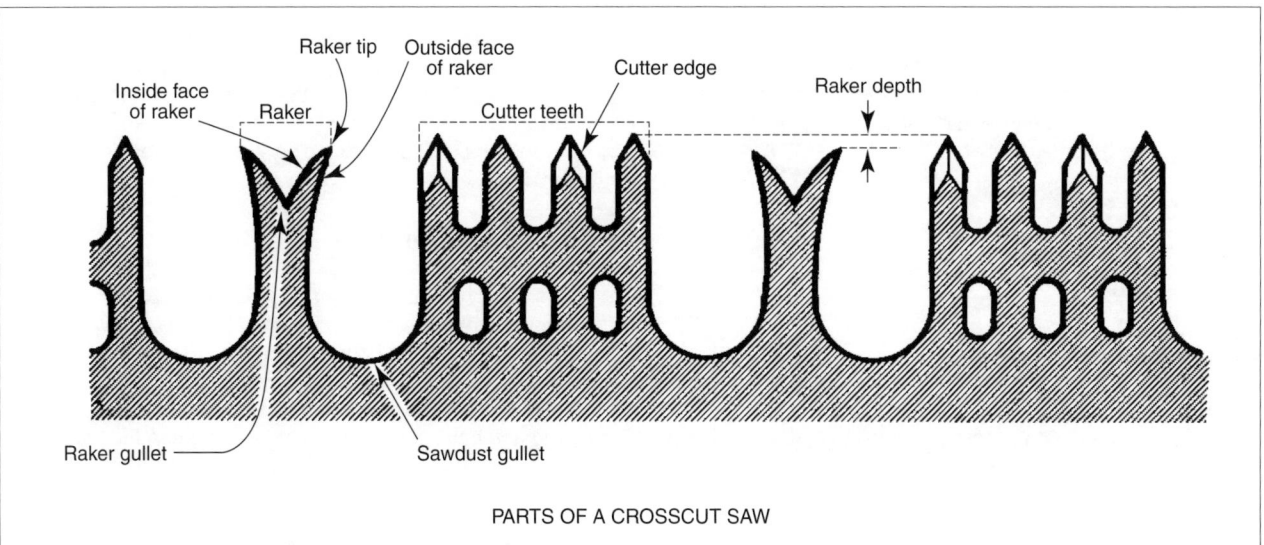

Figure 4-7—Parts of a crosscut saw.

Chapter 4—Crosscut Saw Tasks and Techniques

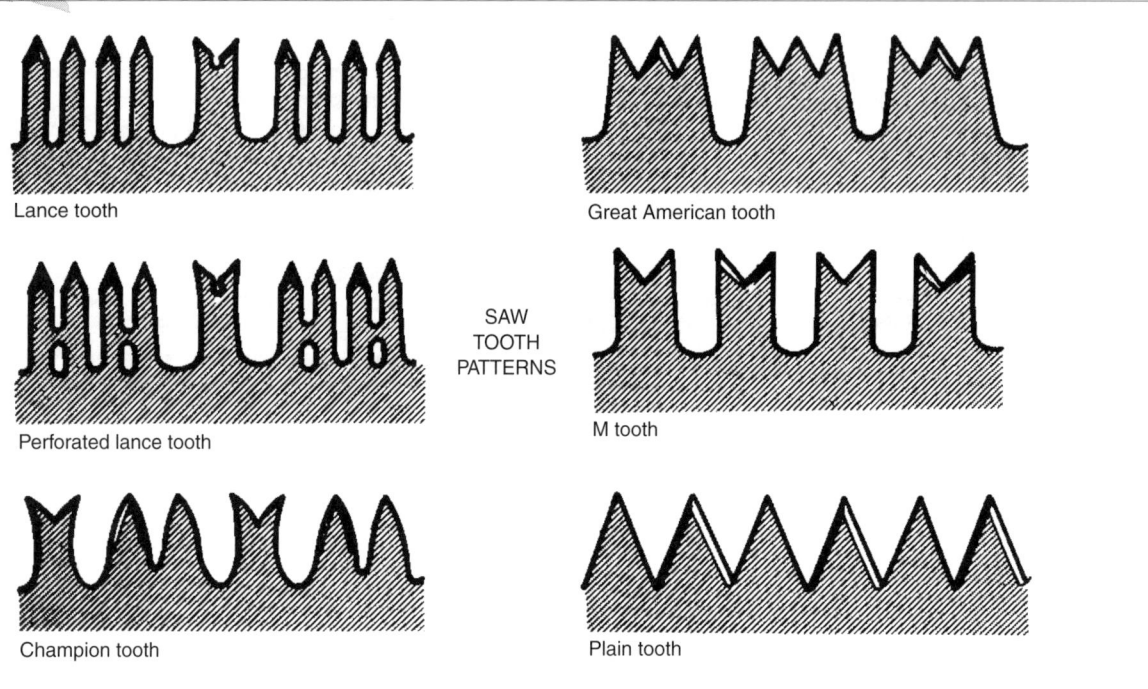

Figure 4-8—Saw tooth patterns.

gullets on each side. The "bridges" between the teeth form the perforations that give the pattern its name. These bridges strengthen the teeth and reduce chatter when the saw is used to cut harder wood.

Lance-Tooth Pattern

For many years the lance-tooth pattern was the standard for felling and bucking timber in the American West. It consists of groups of four alternately set cutters separated by an unset raker with gullets on each side.

Saw Handles

Handle Position

The handle's position on the saw affects the saw's efficiency. Changing either the arm and hand position, or the handle position, changes the delivery of force to the saw.

Handle Attachment Holes

A one-person saw has a fixed D-shaped handle with additional holes on the top of the saw to attach a supplemental handle. Many two-person crosscut saws (usually bucking saws) have two holes on each end for handles. Moving the handle from the lower hole has the same effect as moving the hands several inches up the saw handle. With the handle in the upper hole, a push stroke applies more downward force on the saw, causing the teeth to sink deeper into the wood. The deeper cut requires more force on the pull stroke. A slight upward force is applied to the saw, making it easier to pull.

Types of Saw Handles

Handles may be fastened permanently to the blade with rivets. Removable handles may be fastened to the blade with a steel loop or with a pinned bolt and wingnut assembly.

Quality saw handles are often hard to find. Handles must be strong and must not allow movement between the handle and the blade.

Loop Style: The loop-style handle is a common design. Most of these models have a metal loop running up through a hardwood handle to a nut, which is either inside the handle (plug nut) or part of a cap at the end of the handle. The loop design allows the loop to be slipped over the saw blade. When the wooden handle is turned, the loop tightens around the saw. These models do not use the saw handle holes. Most saws have a notch or a "valley" that the bottom of the loop rests in.

Because these saw blades must have a notch for the loop, they do not have teeth all the way to the end of the blade.

Pin Style: The pin-style handle design—the most common— uses the handle holes in the saw blade.

The climax-style handles were the most common pin-style design. Even today, they appear on some modern two-person crosscut saws.

Perhaps the most common vintage saw handle used today is the Pacific Coast model of the pin-style design. It has a finger guard with a groove to accept the saw blade and two cast flanges that saddle the wooden handle. The $\frac{1}{2}$-inch-diameter rivet pin passes through a hole in the wooden handle. It is secured with a heavy wingnut.

Supplementary handles are used on one-person crosscut saws. The handle can be placed at the end of the saw for an additional sawyer or directly in front of the D-shaped handle when a single sawyer wants to use both hands.

Handle Installation and Maintenance

The wooden handles on crosscut saws are usually select-grade hardwoods $1\frac{1}{4}$ inch in diameter and about 14 inches long. When the handle is not on the saw, it needs to be kept away from sharp edges that could nick or cut it.

Saw Maintenance

The maintenance topics discussed in this section are for the crosscut sawyer. Some topics, such as saw filing, are included just to provide an overview during training. An experienced saw filer should do the filing. The *Crosscut Saw Manual* (technical report 7771-2508-MTDC) by Warren Miller is an excellent resource for more information on saw maintenance and filing.

Cleaning the Saw

Saws need to be clean to function effectively. Clean saws at the end of the day before storing them.

Removing Rust—Rust probably does more damage to saws than anything else. Remove light rust using steel wool. Use a pumice grill block to remove rust that is too heavy to be removed with steel wool. A liberal amount of cleaning solution will keep the block's pores open.

To remove heavier rust, use an ax stone. Always use a liberal amount of cleaning solution. **NEVER** use a dry stone on the saw blade.

As rust and other deposits are removed, you will see imperfections in the saw blade. Spots that are shinier than the rest of the saw are high spots. Spots that are duller than the normal saw surface indicate low spots. A high spot on one side of the blade usually produces a low spot on the other side. These kinks or bends need to be hammered out by an experienced saw filer.

Do not apply too much pressure on the cutter teeth because you can remove metal from the set and reduce tooth length.

A wire brush can be used to remove loose rust and scale. **NEVER** use a power sanding disk on a saw blade.

Removing Pitch—A saw that is well cared for will not rust, but it will develop pitch deposits during normal use. Some pitch can be removed with a citrus-based solvent as the saw is being used by allowing the saw's motion to scrub away the buildup. However, pitch can still be deposited on the saw. Pitch buildups can be removed at the end of the day with steel wool and a cleaning solution.

Using Cleaning Solutions—Limit the use of harsh chemicals for cleaning saws. Wear the proper personal protective equipment and know how to use the cleaning solutions safely. Check the Material Safety Data Sheet if you are unfamiliar with the hazards of using and storing a particular product. A number of citrus-based cleaners on the market are effective and safe.

Naval Gel can be applied to remove heavy rust and scale. Use only as directed, with adequate ventilation. This product stops the chemical reaction of the rust.

Chapter 4—Crosscut Saw Tasks and Techniques

Checking for Straightness

The sawyer should check the saw periodically for straightness. A saw should be checked if it receives any harsh treatment during transportation or use. A saw that is not straight can buckle on the push stroke. The narrower, lighter felling saws are more prone to buckling.

Using Straightedges—Remove the saw handles and hang the saw vertically from one of its handle holes.

Saw filers usually have straightedges made especially for this work. You will need a pair of straightedges. Two combination square rules can also be used. Before using the straightedges on the saw, hold them together and make sure they maintain contact along their entire length. You should not see light between them when you put them together and hold them up to a light source.

Straightedges work by allowing you to feel the difference in resistance between the saw and the straightedge as the straightedges are twisted back and forth over the saw's surface. The straightedges are moved as a pair with the saw between them. You will feel increased drag on the ends of the straightedge on the side of a saw with a depression. On the other side of the saw, the straightedge will pivot easily on the corresponding bump. Even resistance on both straightedges reflects a straight saw that does not have any kinks, bends, or bumps.

If you find any major irregularities, report them to the person who files your saws.

Testing the Saw

Testing determines whether a saw cuts straight, runs smooth, and produces long, thick shavings. The saw should produce shavings and not sawdust. The longer and more abundant the shavings, the better the saw is performing. Green logs produce longer shavings than dry logs. The shavings should be long and thick with smooth edges. If the edges of the shavings have "whiskers" or irregularities, the rakers are probably too long. If the shavings are paper thin, the rakers are too short (figure 4-9).

Does the Saw Cut Straight? Cut far enough into the test log to determine if the cut is perfectly straight. If the saw consistently pulls to one side through no fault of the sawyer, the saw needs additional maintenance.

Sometimes a sawyer standing in an awkward position can put a twist or bend on the saw. A saw will not cut straight if it is kinked or bent.

Too much set on one side of the cutters can cause the saw to pull to that side. If a saw has been sharpened improperly, the teeth may be longer on one side than the other. The saw will pull to the side with the longer teeth.

NEVER field sharpen or "touch up" dull cutters. Doing so shortens the teeth, compounding the problem.

Does the Saw Run Smooth? Look for a saw that does not chatter or seem like it is jumping through the log. The saw may also feel like it alternately catches and releases. A smooth-running saw seems to cut effortlessly.

Smoothness is most associated with the rakers. If a saw feels like it is snagging the wood, it is probably because one or more rakers have been filed incorrectly.

Inconsistent set in the teeth can also produce a jumpy saw. Look at the walls of the cut. The cut surfaces should be smooth.

A sawyer cannot do anything to fix a saw that is running rough. A qualified saw filer will need to make the necessary adjustments.

Brief Overview of Saw Filing Procedures

This overview will not teach a student to file a crosscut saw. But it will allow the student to understand the skill and labor required to sharpen and recondition a crosscut saw. Specialized tools are needed to file saws. Filing must be done by a qualified filer in a saw shop.

The crosscut saw may be the most precise tool that a woods worker uses. An experienced filer setting teeth can feel the

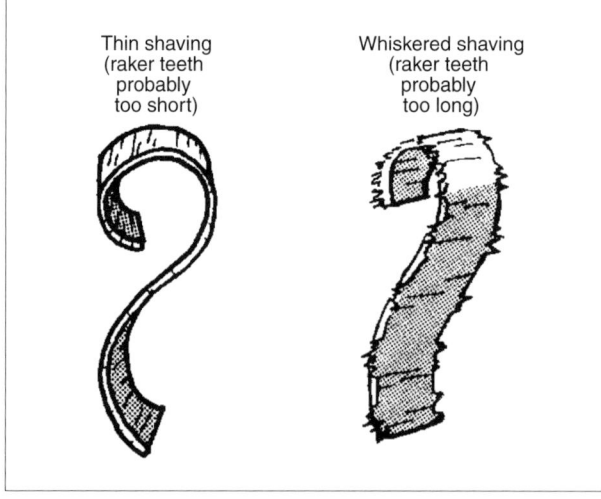

Figure 4-9—Examine the saw shaving.

difference between a 0.013-inch and a 0.012-inch set. That means the filer is making a 0.0005-inch judgment.

Saw Vises and Tools—A filer needs to work in a well-lighted location with a wooden vise to hold the saw.

Straightening—Straightening is an art in itself. The filer must carefully move the metal by hammering the blade on an anvil.

Jointing—After the saw has been cleaned and straightened, jointing is the first step in sharpening. A tool called a jointer holds the file. The points are filed off the cutter tips so that each of them lies on the circle of the saw (figure 4-10).

Fitting Rakers—The raker gullet is shaped using a triangular file. The raker is lowered and checked with a pin gauge, which establishes the exact clearance below the cutters.

Tooth Pointing—Each tooth is sharpened to a point. The filer has the option to make the bevel suit the wood type.

Setting Teeth—The teeth need to be set so they lie directly behind one another. The filer puts equal set in all the teeth by hammering the point over a beveled hand anvil. The set is checked using a tool called a spider.

Storage

Whether they are stored at a backcountry guard station or at a unit's warehouse, crosscut saws need to be stored properly.

Long-Term Storage—Store crosscut saws straight. Remove the handles and store the saws in a dry location.

NEVER store a saw flat on a metal surface. It is best to hang a saw from a nail through a handle hole. Although the saw can be laid horizontally if it is supported along its entire length, items may be dropped on a saw, damaging it. During long-term storage, oil will bleed into the saw's wooden handle if the saw is lying flat.

Apply canola oil or another environmentally sensitive lubricant before storing a saw. Wear appropriate gloves when applying the oil.

NEVER lean a saw against a wall where it could develop a bend.

NEVER leave a saw bent around a fire pack.

DO NOT store a saw in a sheath or with a guard on. Rubber-lined fire hose is particularly bad because it traps moisture, holding the moisture next to the saw's teeth.

DO NOT hang a saw where animals or people could be injured by the unsheathed teeth.

DO NOT store saws on top of one another. When the unsheathed saws rub against each other the saws can be damaged.

In the Field—Saws need to be wiped clean and rubbed with canola oil or another environmentally sensitive lubricant before you leave them in the field. Choose a storage location out of human sight and away from game trails. If saws are only being left overnight, they can be laid under a log with the teeth pointed in.

Remove the saw handles and sheaths. Bears tend to gnaw on wooden handles. Rodents chew on leather straps and anything that has salt on it. Leave nothing but the metal parts in the field. If you are storing a saw longer than for just one night, hang it.

Saw Sheaths—Sheaths protect the saw and prevent it from causing damage or inflicting injury. Saws should be sheathed as much as possible unless they are being used or are in storage. Wear gloves when removing or replacing a saw sheath.

A length of old firehose that has been split makes one of the best crosscut saw sheaths. Wipe the hose's rubber inner lining with an oily rag to repel water and reduce the possibility that moisture in the sheath will cause the saw to rust.

Attach the firehose to the saw using parachute cord or Velcro closures. To install the hose sheath, begin by rolling it inside out (rubber side out). Turn the saw so its teeth face up; unroll the hose down the saw, covering the teeth.

Some sawyers sandwich the saw between two rectangular pieces of plywood. The saw's handle holes are placed over pins at each end of one of the pieces of plywood, securing the saw.

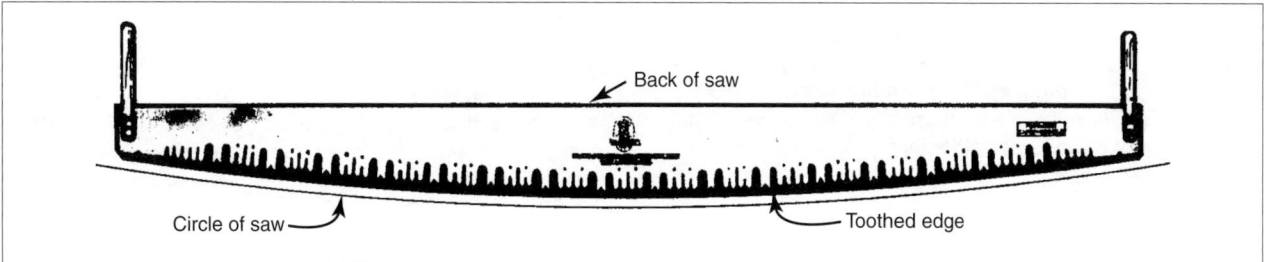

Figure 4-10—Circle of the saw.

Transporting Saws

Saws must be transported so they will not be damaged, so they will not injure people or livestock, and so they will not damage property and equipment.

Saws are difficult to transport because they are long and flexible. Vintage saws are bent to make them easier for hikers or packstock to carry. Saws can be delivered by parachute. Modern saws **SHOULD NOT** be bent. The softer metal will hold the bend.

Because saws may be taken by boat, plane, helicopter, truck, dog sled, or packstock, or be carried by a hiker during different legs of a journey, several types of protection may be needed to get a saw to the work site. Sheaths should always cover saws when they are being transported.

Saws get hot in the sun. Use gloves to handle a saw that has been lying in the sun.

Boats—If a saw is being transported in an open skiff, remove the saw's handles and place the sheathed saw on top of the other cargo. On many boats the place that is the most out of the way is along the gunwales. Open boats can take on a lot of spray. Saltwater spray can cause rust. **ALWAYS** coat the saw with canola oil or another environmentally sensitive lubricant before transporting it. Once the saw is on land, remove the sheath and rinse off any salt with a good freshwater bath.

Saws transported on kayaks are best secured to the bow where they can be seen. In canoes, carry saws in the center on the floorboard. Transport the saw without handles in a rubber-lined hose sheath. Secure the saw by tying parachute cord through the handle holes. Be sure to remove the sheath and dry the saw after arriving at your destination.

Aircraft—In small aircraft, the handles can get in the way. If a saw's handles must be removed to bend the saw into a loop, wire **MUST** be strung through the handle holes to secure the saws. **DO NOT** string parachute cord or any other nonmetallic material (including nylon ties) through the handle holes to secure the saw. Jostling during the flight could cut nonmetallic materials, allowing the saw to spring to full length. The same considerations apply when saws are carried in helicopters.

Helicopters can transport saws as an external load. Use care when packing saws that are carried as sling loads by helicopters. One way to reduce breakage is to carefully bend the saw around a box. Place the box in the middle of the sling bag with the saw's ends down. Stack other materials around the saw.

Vehicles—When transporting crosscut saws in a pickup truck, lay the sheathed saw flat on the bed of the truck. Don't place heavy tools on top of the saw.

Dog Sleds—In some areas saws are transported by dog sled. If you do not expect to use the saw to clear trails during the trip, sheathe the saw and place it on the bottom of the sled. If the saw may be needed for trail work, place it along one side of the sled where it will be easier to reach.

Packstock—Take extra care when carrying a crosscut saw on packstock. Select the gentlest animal to carry the saw. Put that animal in the lead where you can easily see the saw.

The handles on a two-person saw shall remain attached. Sheathe the saw with firehose and wrap the saw in a mantie with the handles exposed and secure. Bend and place the saw over the animal with the teeth facing to the rear. Tie the saw down to the latigo or double cinch. One-person saws can be transported on riding stock in a leather or canvas scabbard (similar to a rifle scabbard). A piece of hardwood protects the scabbard from the saw's teeth.

Hikers—Saws should be sheathed when you are hiking to the job site. The person carrying the saw should be the last person in line.

Two-person saws should have the rear handle removed. If the handle is left on, it can snag on branches.

The saw can be carried on your shoulder with the teeth facing outward. **AVOID** carrying the saw with the teeth pointing upward. Carry the saw on your downhill shoulder so you can throw it off if you slip or fall.

A vintage saw can be bent around a pack if it is being carried for long distances. Usually both handles are left on to secure the saw in its bent position.

Saw-Related Tools and Equipment

After completing this section, students will:

- Understand the importance of careful selection of tools for crosscut saws.
- Have a working knowledge of the use of wedges in crosscut saw applications and how the use of wedges differs when cutting with a crosscut saw rather than a chain saw.

Lubricants

Types

Water-based lubricants (often including citrus-based ingredients) and petroleum-free lubricants (based on canola oil) are available commercially.

Functions

Saw teeth do not need to be lubricated as they cut. The friction of the saw teeth set against the kerf keeps the teeth reasonably clean. However, resin deposits on the lower part of the teeth and in the saw gullets produce drag. Lubricants can soften these deposits and help remove them.

Cutting in extremely wet environments or during a hard rain can cause wood fibers to swell. In these conditions, an oil-based lubricant can help reduce drag. At the end of each day, clean the saw with a solvent and apply a thin coat of oil.

Applying Lubricants

Open containers waste lubricants. Squeeze bottles allow the sawyer to direct a stream of lubricant onto the saw's surface. On the pull stroke, the sawyer keeps one hand on the saw handle and applies the lubricant with the other, putting the bottle down before the next push stroke.

Axes

Axes need to be heavy enough (3 to 5 pounds) to drive wedges into the trees being felled. The back of the ax should be smooth, have rounded edges, and be free of burrs to minimize damage to wedges. Pulaskis should never be used to drive wedges.

Always remove branches, underbrush, overhead obstructions, or debris that might interfere with limbing and chopping. Do not allow anyone to stand in the immediate area. Make sure workers know how far materials may fly. Protect all workers against flying chips and other chopping hazards by requiring them to wear the appropriate PPE.

Always position your body securely while working with a tool. Never chop crosshanded; always use a natural striking action. Be alert when working on hillsides or uneven ground. If you cut a sapling that is held down by a fallen log, the sapling may spring back. Be alert for sudden breakage. If you do not have a need to cut something, leave it alone.

Never use chopping tools as wedges or mauls. Do not allow two persons to chop or drive wedges together on the same tree. When chopping limbs from a felled tree, stand on the opposite side of the log from the limb being chopped and swing toward the top of the tree or branch. Do not allow the tool handle to drop below a plane that is parallel with the ground unless you are chopping on the side of a tree opposite your body.

If the cutting edge picks up a wood chip, stop. Remove the chip before continuing. To prevent blows from glancing, keep the striking angle of the tool head perpendicular to the tree trunk.

Wedges

Wedges are essential tools for safe felling and bucking. They provide a way to lift the tree, preventing the tree from sitting back when it is being felled. A wedge must be inserted into the backcut as soon as possible. Wedges also reduce binds on the saw when bucking.

Select the correct wedge for the job. The proper type, size, and length or a wedge varies, depending on its use. The size of the tree being felled or the material being bucked determines the size of the wedge that will be needed. If the wedge is too small, it may be ineffective. If the wedge is too long, it may not be able to do its job without being driven so far into the tree that it contacts the chain.

Always drive wedges by striking them squarely on the head. Drive them carefully to prevent them from flying out of the cut.

Check wedges daily or before each job. Do not use cracked or flawed wedges. Wedges that are damaged need to be cleaned up before they are used again.

Recondition heads and the tapered ends when grinding wedges to the manufacturer's original shape and angle. Wear eye protection and a dust mask.

Repair any driving tool or remove it from service when its head begins to chip or mushroom.

Carry wedges in an appropriate belt or other container, not in the pockets of clothing.

Most wedges are made out of plastic or soft metal, such as magnesium, and come in different sizes. Use plastic wedges in both felling and bucking operations to prevent damaging the saw if it contacts the wedges.

The two basic types of wedges used in sawing are single (figure 4-11) and double (figure 4-12) taper.

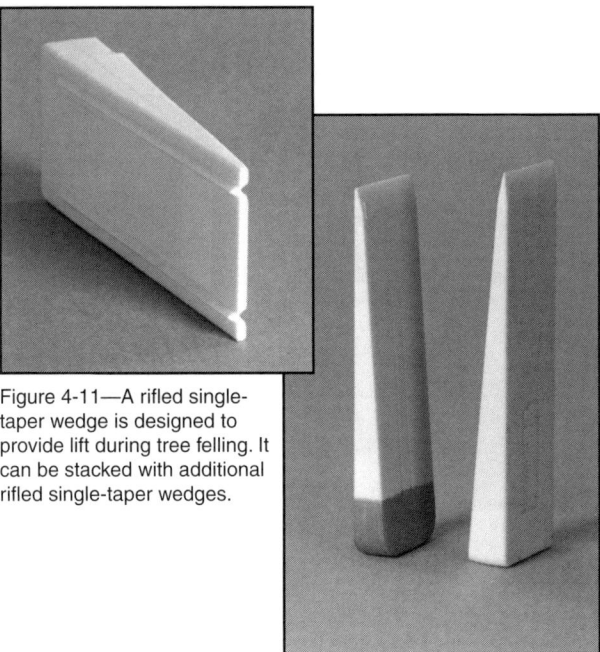

Figure 4-11—A rifled single-taper wedge is designed to provide lift during tree felling. It can be stacked with additional rifled single-taper wedges.

Figure 4-12—Double-taper wedges are designed to reduce bind.

Splitting Wedges

Splitting wedges are always made of steel. Magnesium wedges should not be used to split wood. Steel wedges with smooth faces sometimes rebound out of cuts when they are driven. Most steel wedges have shallow grooves or depressions below the wedge face. The backward motion of the wedge is reduced as wood fills these voids.

Smooth faces can be roughened up by a cold chisel. Wedges must be firmly set before they are driven with a double jack.

Lifting Wedges

Lifting wedges are tapered on just one face. They are truly an inclined plane. Wedges to reduce bind or split wood are double-tapered, meaning that each of the broad faces taper equally from the center. When such wedges are driven, the force is equal on both sides, causing the wood to move equally.

Lifting wedges exert force in the direction of the inclined plane. Two wedges can be stacked one on top of the other to produce an even lift. Lifting wedges have many applications. They can be used to tighten, pry apart, or move materials. Exerting a force in one direction can be valuable. The sawyer may need to exert a force in one direction when getting a saw unstuck or when removing a chunk of log if a carelessly placed compound cut gets bound up.

Plastic or steel wedges can be bought with a single taper. These wedges often have a groove on the sole face to increase holding power on that side. The lifting or moving takes place on the smooth side that serves as the inclined plane.

Peaveys and Cant Hooks

The blacksmith Joseph Peavey invented the peavey. Both the peavey and the cant hook use a curved metal hook on the end of a straight handle to roll or skid logs. A peavey has a sharp pointed spike at the lower end, while a cant hook has a tow or lip. Most peaveys and cant hooks come with a duckbill hook that is a good all-around style. Peaveys and cant hooks come with hickory handles that are from 2 to 5½ feet long.

Peaveys are used almost exclusively in the woods, where the pick is used for prying. Peaveys are handy for prying logs up onto blocks to keep the saw from pinching while bucking. The cant hook is used primarily to roll logs.

- Keep the handles free of splinters, splits, and cracks.
- Keep points sharp.
- Keep your body balanced when pushing or pulling the pole.
- Grip the handle firmly; do not overstress it.
- Place a guard on the point when the tool is not in use.

Underbucks

Underbucks help hold the saw in position when the saw is cutting from underneath the log. They also act as a fulcrum. A good sawyer can cut as fast—or faster—from underneath a log as from the top. When the sawyer applies a downward pressure on the handle, the saw is forced up into the log. The sawyer does not have that mechanical advantage when cutting from the top. We will discuss several types of underbucks.

Types of Underbucks

Axes are the most common type of underbuck. Mechanical underbucks (figure 4-13) are sometimes used instead of an ax. Axes used for underbucking should have a 36-inch wooden handle that has been slightly modified. Cut two series of three notches on one side of the handle about 6 inches from the end. This allows room for your gripping hand when you use the ax for chopping. The series of three notches, placed about an inch apart, allows the sawyer to more accurately line up one of the notches with the cut. The notches should be 30 to 45 degrees off perpendicular to allow room for the saw between the ax handle and the log.

Figure 4-13—A mechanical underbuck helps hold the saw up when cutting from the underside of the log.

Bucking and Felling Preparation and Techniques

After completing this section, students will:

- Understand proper preparation for bucking and felling.
- Know the hazards and binds associated with bucking and felling operations. Have a working knowledge of the different types of cuts needed to use a crosscut saw for bucking and felling.

Bucking

Safety Considerations

The same principles apply whether a crosscut saw or a chain saw is used for bucking or felling, but the sawyer is exposed to risks longer during crosscut saw operations. Great care needs to be taken when bucking or felling.

Situational Awareness for Bucking—Plan the bucking cut carefully after considering:

- The escape route.
- Slope.
- Tension.
- Compression.
- Rocks and foreign objects on the log.
- Pivot points.
- Adequate saw clearance.
- Overhead hazards.
- The limits of your ability.
- The length of the guide bar in relation to the log being bucked.
- People and property in the cutting zone.
- Spring poles.
- Proper tool placement.
- Falling or rolling root wads.
- The log's tendency to roll, slide, or bind.
- Broken-off limbs underneath the log that can hook the sawyer if the log rolls.
- The footing.

Bucking Sizeup

Spring Poles—Spring poles are limbs or saplings that are bent under a fallen tree. These poles can store tremendous amounts of energy. Spring poles can be dangerous if they are cut accidentally, or without careful planning. Cut a spring pole only when necessary.

First, determine what will happen when the spring pole is cut. The cut needs to be made from a safe location. A crosscut saw is not used to release a spring pole unless the pole is very large. Normally an ax, pruning saw, or pulaski is used. Spring poles are under extreme compression and tension. The generally accepted way to remove a spring pole is to make a series of small cuts on the side under tension. Cuts need to be slow, allowing time for the wood to respond to the changing forces.

Suspended Logs—Cutting a suspended log is a single-buck operation. Often only one side is safe or has adequate footing for you to make the cut.

If you are standing on blowdown where several trees are jackstrawed in different directions, carefully evaluate the sequence in which trees should be removed. Generally, **REMOVE THE BOTTOM LOGS FIRST**. This practice reduces the chance that top logs or other material will move.

It might not be possible to remove all suspended trees with a saw. Only take out the ones that can be removed safely. Other suspended trees can be removed with winches or explosives, if necessary.

Chapter 4—Crosscut Saw Tasks and Techniques

Suspended logs often roll when they are released. Be sure the log has a safe path to travel. Logs may ricochet off other objects, making their path unpredictable. **BE SURE** no snags or other weak trees are in the log's path. They could snap if they were struck by a rolling log. Fell snags or weak trees first, if they can be felled safely.

Unsound Wood—Unsound wood can crack or break without warning. It can be hazardous because it is unpredictable. Logs may be sound in one area and rotten in other areas. Examine ends of logs and look for indications of rot. Observe the color of shavings the saw is producing. Dark shavings indicate rot. Rotten wood doesn't hold wedges well, making them ineffective. Because rotten logs may hold more moisture, saws tend to "load up", increasing the need to use wedges to keep the kerf open.

Planning the Cut

Can the log be safely bucked with existing skills and equipment?

Sawyers should not feel pressured to perform any task that is above their ability. Ask other crew members to silently sizeup the situation. Discuss findings afterward.

Types of Cuts—The three basic types of cuts are: the straight cut, compound cut, and the offset cut. We will describe each type of cut and its usual application (figure 4-14).

A straight cut is made through the log from one side. It can be performed by single or double bucking. It can also be cut from underneath the log by a single sawyer (using an underbuck).

A compound cut is placed at an angle less than perpendicular to the log and angled so that the bottom of the cut slopes toward the part of the log that is being removed. This cut is typically used when clearing a large log that is across a trail. Two cuts need to be made and the severed chunk of the log has to be removed.

The offset cut is placed so that the bottom underbucking cut **DOES NOT** match up exactly with the top cut. This kind of underbucking operation is used when a log is suspended and will drop free when severed. Once the top cut has been made, a single sawyer selects a groove (about one-half inch toward the ax head from the top kerf) from the grooves cut into the saw handle or installs a mechanical underbuck. The ax head is always secured to the side of the log that won't move when the log is cut.

This small amount of offset wood acts like the holding wood left when trees are felled. In felling, the holding wood keeps the tree from kicking back. In underbucking, the offset wood prevents the severed log from damaging the saw when the log drops. If the offset wood is severed, control is lost. If the cuts meet, the log will want to carry the saw with it when the log drops. Because the ax handle supports the saw, the saw's force can break the ax handle. The saw may fly upward, possibly injuring the sawyer, or bending, kinking, or snapping the saw.

Determining Binds—Understanding directional pressures, or binds, is important. These binds determine bucking techniques and procedures.

Landforms, stumps, blowdown, and other obstacles that prevent a log from lying flat cause binds. Binds produce different pressure areas (figure 4-15). The tension area is the portion of the log where the wood fibers are being stretched apart. In this portion of the log, the saw's cut (kerf) opens as the cut is made. In the compression area, the wood fibers push together. In this portion of the log, the kerf closes as the cut is made.

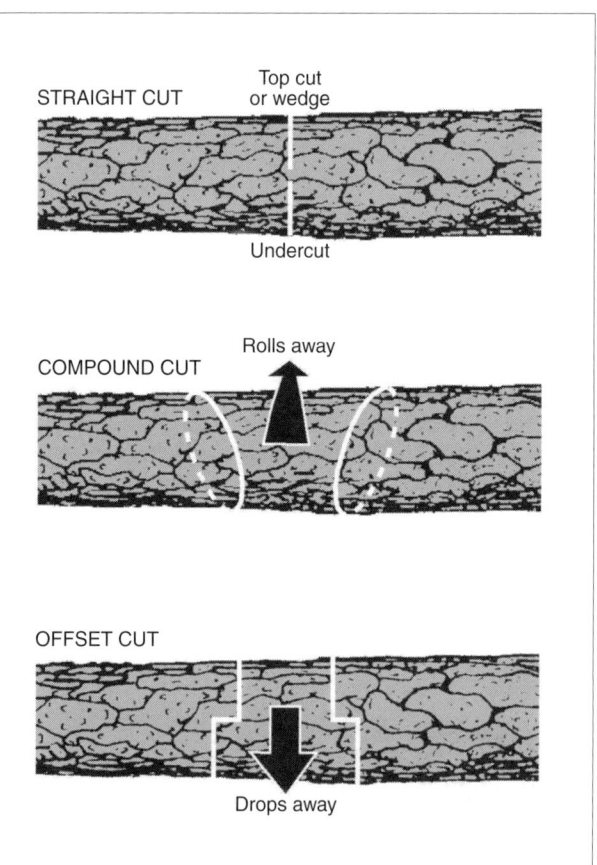

Figure 4-14—Three basic cuts: straight, compound, and offset.

Chapter 4—Crosscut Saw Tasks and Techniques

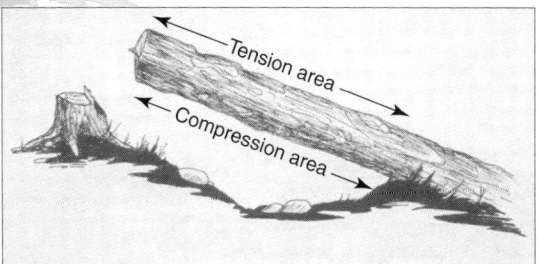

Figure 4-15—When a bind occurs, pressure areas result. These areas are called tension (pulling apart) and compression (pushing together) areas.

It is extremely important to determine what will happen to the log when it is cut. Inspect the log for all binds, pivot points, and natural skids. Various bucking techniques can be used to lower a suspended tree to the ground.

The four types of bind are: top, bottom, side, and end (figure 4-16). Normally logs have a combination of two or more binds:

Top Bind: The tension area is on the bottom of the log. The compression area is on the top.

Bottom Bind: The tension area is on the top of the log. The compression area is on the bottom.

Side Bind: Pressure is exerted sideways on the log.

End Bind: Weight causes compression on the log's entire cross section.

Determine Bucking Locations—It is best to start bucking at the top of the log and work toward the butt end, removing the binds in smaller material first. Look for broken limbs and tops above the working area. Never stand under an overhead hazard while bucking.

Look for spring poles (figure 4-17). Look for small trees and limbs bent under the log being bucked. They may spring up as the log rolls away. If you can safely do so, cut them off before the log is bucked. Otherwise, move to a new cutting location and flag the hazard. Anticipate the spring poles' reactions.

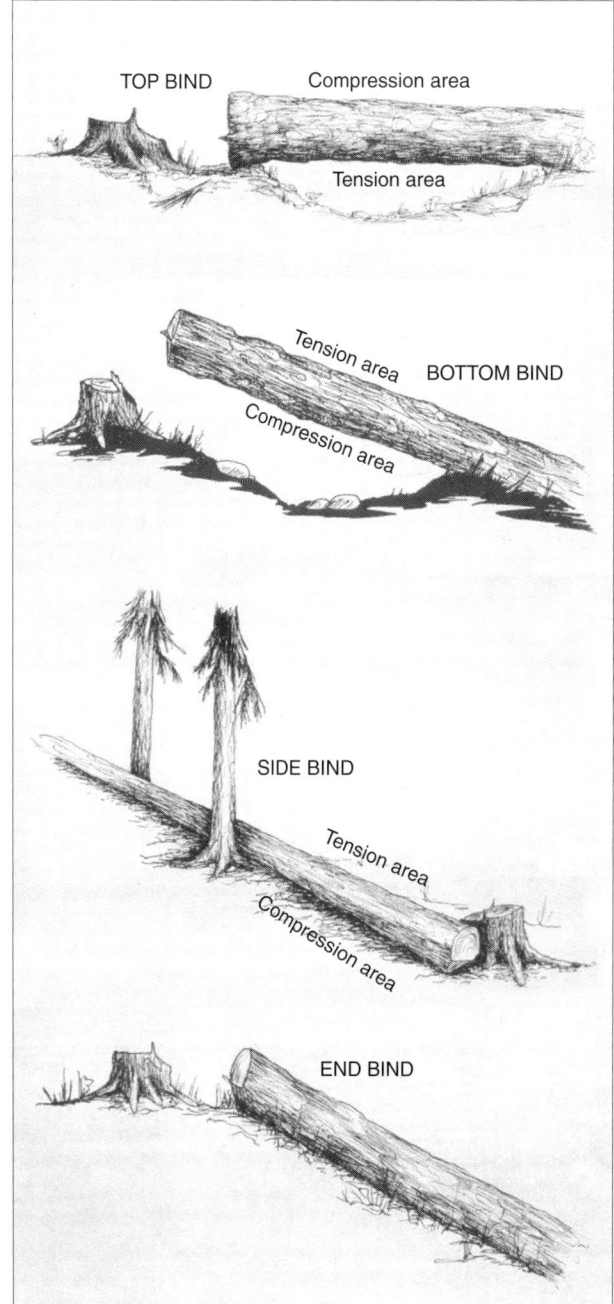

Figure 4-16–There are four types of binds. A log can have a combination of two or more binds.

57

Figure 4-17—Look for spring poles. They can release and cause accidents.

Figure 4-19—Watch the kerf for movement that will indicate a bottom bind (kerf opens) or top bind (kerf closes).

Determine the offside (figure 4-18). The offside is the side the log might move to when it is cut, normally the downhill side. Watch out for possible pivots. Clear the work area and escape route. Allow room, more than 8 feet, to escape when the final cut is made. Establish solid footing and remove debris that may hinder your escape.

Cut the offside first. If possible, make a cut about one-third the diameter of log. This allows the sawyer to step back from the log on the final cut.

Watch the kerf to detect log movement. Position yourself so you can detect a slight opening or closing of the kerf. There is no better indicator of the log's reaction on the release cut. If the bind cannot be evaluated, proceed with caution. It may be necessary to move the saw back and forth slowly in the kerf to prevent the saw from getting bound as the cut pressure closes the kerf. Cut only far enough to place a wedge. Continue cutting. Watch the kerf. If the kerf starts to open, the log has a bottom bind. If the kerf starts to close, the log has a top bind (figure 4-19).

Reduce the remaining wood. Visually project the kerf's location to the bottom of the log. Reduce the amount of wood to cut on final cut by cutting a short distance into the log along this line. Be prepared for kickback.

Hazards of Bucking in Blowdown—Blowdown is a result of strong winds that have uprooted the trees. At any time while the bucking cuts are made, the roots can drop back into place or roll in any direction. Avoid standing directly behind or downhill from the roots.

Small trees growing on the roots of blowdown (figure 4-20) could be forced into the sawyer's position if the roots drop or roll. Cut the small trees off first. Limbs may be preventing the roots from rolling. Don't cut off those limbs.

Figure 4-18—Determine and stay clear of the offside (downhill side) when you are bucking.

Figure 4-20—Small trees growing from the roots of blowdown can move if the roots drop or roll.

Chapter 4—Crosscut Saw Tasks and Techniques

Points to Remember...

- Do a complete sizeup. Identify the hazards and establish your escape routes and safety zones.

- Use objects such as rocks, stumps (if they are tall enough), and sound standing trees with no overhead hazards for protection in the event the tree springs sideways toward you when you make the release.

- Removing a single section of log may require that other binds be eliminated first. Angled bucking cuts, wide on top and made on the offside, allow a single section of log to be removed. Angled cuts will permit the bucked section of log to be rolled away from the rest of the log. Buck small sections that will be easy to control when they begin moving.

- Binds and log movement will change. Reevaluate as necessary.

- Warn workers who are working in and below an active cutting area. Allow workers time to move to a safe location. Verify their safety visually and verbally. Announce when a bucking operation has been completed.

- All logs must be completely severed when they are bucked. Use flagging to mark an incompletely bucked log, as a hazard. Never approach a cutting operation from below.

Single-Bucking Techniques

New sawyers should master the skill of single bucking before learning double bucking. If new sawyers can handle a long two-person saw alone, they have mastered the principles of keeping the saw running smoothly without buckling. Thinner, lighter felling saws are hard to use for single bucking except by very experienced sawyers. The stiffer, heavier bucking saw is easier to push during single-bucking.

The reasons to single buck are:

- The sawyer starts out double bucking and needs to finish the cut from one side because of safety considerations or log movement.

- The log is too large for the length of the saw.

- The sawing sequence starts or ends with underbucking, which can be done only by a single sawyer.

As a general rule, the saw needs to be twice as long as the log's diameter plus 6 inches. Imagine trying to cut a 4-foot-diameter log with a 7-foot saw. If the center raker and adjoining gullets are in the center of the log, each sawyer needs at least 2 feet of free blade so the center gullets will clear the log. With a 7-foot saw, the shavings would never be removed from the gullets in the center foot of the saw. On each stroke, the gullets pick up more shavings. As the gullets fill, the saw works harder and binds, especially if the wood is green and pitchy. Usually sawyers can't run the saw right to the handles and still protect their hands from being drawn into the bark.

A single sawyer can take off the handle at one end of the saw. That end of the saw can be drawn into the log, allowing the shavings to be removed from the gullets.

When making compound cuts, the length of the cut needs to be used to determine how long the saw needs to be. The saw does not work as efficiently in a sloping cut or a compound cut as it does in a crosscut. The more angle that is placed on a compound cut, the less effective the saw is working. Compound cuts can make for some hard sawing.

Single Bucking With No Bind: Top Cutting

• Lay the unsheathed saw on its side over the log to be bucked. Sprinkle lubricant on both sides of the saw.

• Hold the saw in the dominant hand and guide the back of the saw with the other hand for a few strokes until the saw is set in the kerf.

• Insert wedges as soon as possible, driving them snug. Take care not to hit the saw.

• Lubricate the blade as needed just before the push stroke. On a smaller log, the sawyer may be able to lubricate the far side of the saw just before the pull stroke. Be sure to lubricate both sides of the saw blade equally.

• As the cut is ending, use only the teeth at the end of the saw blade. This technique prevents the log from damaging the "production" cutters near the center of the saw when the log rolls or pinches the saw.

Single Bucking With Top Bind: Underbucking Required

Underbucking is used when the log has a top bind and you can get under the log. The first cut must be started from the top because the top of the log is under compression. If the compression is not corrected, the kerf may close and pinch the saw.

After you have inserted the wedges and driven them snug, continue cutting down from the top, leaving enough uncut wood to support the log's weight. Because the top of the log is under compression, the bottom is under tension. The more compression you relieve, the greater the tension on the bottom of the log. The log will start to equalize this pressure by exerting pressure on the wedges. If you use two or more wedges spaced at the 10 and 2 o'clock positions, you can spread the force over a large area. If only one wedge is used at the 12 o'clock position, all the energy is directed to that relatively small area.

Remove the saw from the top cut and prepare to finish the cut from the bottom by underbucking. A log or rock can be placed under one side of the cut to support the log so it will be less likely to carry the saw to the ground when the cut is completed.

Underbucking

During sizeup, you determined which side of the severed log will probably remain the most stationary, providing the anchor point for the underbuck. A common mistake is to place the underbuck on the side that it is easiest to reach. If this side of the log moves when the log is severed, the saw could be damaged.

To underbuck, use a mechanical underbuck or plant an ax in the log so the handle can be used as a support for the back of the saw (figure 4-21). Line up the underbuck grooves in the ax handle with the top saw kerf and forcefully swing the ax into the log.

Figure 4-21—An ax planted in the lower part of the log can work as an underbuck.

Oil in the underbuck groove will help the saw run easily and will reduce wear on the ax handle. Adjust the handle angle to allow room for the saw to be inserted and for the underbuck to be parallel to the saw kerf.

If you are placing an offset cut, allow for about one-half inch of offset toward the ax head.

If you are underbucking a compound cut, try to have the cuts match exactly because an offset could prevent the log from being freed. Several more wedges may have to be placed in the top cut to provide additional bearing pressure on the kerf faces, holding the log in place.

Lubricate both sides of the saw and the ax handle grooves.

Your body position will determine how to position the ax handle. Usually the handle is reversed so the longer side of the handle is facing up. This allows better delivery of the arm's energy to the saw's teeth.

Place the back of the inverted saw in the underbuck groove. The saw typically starts out at an angle of about 45 degrees from horizontal. Your guiding hand holds the back of the saw. With a light downward pressure on the underbuck, push the saw forward. Pressure on the underbuck needs to be consistent on the push and pull strokes.

After several strokes, you can remove your hand and continue normal cutting. With continued downward pressure, the end of the saw will be doing more of the cutting and the saw blade will level out. As the cut nears completion, be prepared for the severed log to drop.

Single Bucking With Top Bind: Top Cutting

Several methods can be used to buck a log when there is top bind and not enough room to get the saw under the log for underbucking.

All sawing will be from the top. Do a good job of wedging to keep the kerf open. Follow the instructions for wedging (chapter 2). Periodically, drive all the wedges until they are snug. Do not allow wedges to contact the saw.

The cut will want to open up at the bottom. A log or other material can be placed under the log segment that will drop when the cut has been completed, reducing the distance a severed log segment will fall.

Single Bucking With Bottom Bind: Top Cutting

When there is bottom bind and not enough room to get the saw under the log for an undercut, all the cutting will be done from the top. The main problem with bottom binds is that standard wedging does not help. In addition, when the log is severed, segments of the log may drop or roll.

Cut the log as explained for top cutting with single bucking. Lightly place a small plastic wedge at the top of the cut. Do not drive the wedge in.

This wedge will show when the kerf begins to open. When the kerf opens, drive two fan-shaped metal wedges across the kerf. The point is to slow the opening of the kerf and the settling of the log.

As the kerf opens at the top, it exerts more and more compression on the uncut wood. If the force becomes too great, the uncut wood may slab off, possibly damaging the saw. The saw could be damaged even if the log does not slab.

When the log is sawn through, the log's weight may pull out the metal wedges, causing the log to drop and roll.

The sawyer needs to keep sawing or even speed up sawing to keep opening the kerf and relieving the compression pressure. The combination of the using wedges to slow the kerf's opening and speeding up the cutting can prevent additional binding.

Perhaps the best technique to reduce the effects of a bottom bind is to insert a stick into the opening saw kerf. A straight, finger-sized limb about a foot long can be inserted into the opening kerf at the top of the log (do not use plastic wedges).

As the kerf continues to open, the stick slides into the kerf. The stick does not drop to the back of the saw because it is too thick. As the cut is completed, the two halves of the severed log hinge on the stick. The bottom opens up, allowing the saw to drop free.

Single Bucking With Bottom Bind: Underbucking

If you have bottom bind and can get under the log, make the first cut from the bottom. In this case, wedging is not as critical. When the first cut is from the top, fan-shaped metal wedges can reduce the speed at which the kerf opens. A stick can be inserted into the opening top cut if the log is large enough for the stick to fit into the kerf.

Single Bucking With End Bind

If more than one cut is being made, make the top cut first or make the cut where the log has the least amount of weight above the cut. This reduces the end bind on the second cut.

If you are cutting down directly from the top, use more plastic wedges around the cut, especially as it progresses below the centerline of the log. This reduces the possibility of binding.

Single Bucking With Side Bind

This is one of the most difficult and hazardous binding situations.

If there is room below the log for the saw's end to clear, cut the side with compression wood first. The finish cut is on the side with tension wood. Alternately saw and chop out wood with an ax. The saw should be in a nearly vertical position. Always find a safe position to make the finish cut.

If the log is on the ground in a side-bind situation, options are limited. For trees larger than 20 inches d.b.h., the only options are to place a cut beyond the side bind area or to cut out the area with an ax.

Double-Bucking Techniques

New sawyers should master the skills of single bucking before learning double bucking. The reasons to double buck are:

- Large logs can be sawed more easily by two sawyers.

- Two sawyers can transport equipment more easily than one.

Attach both saw handles before removing the sheath. After the sheath has been removed, the uphill sawyer normally hands the saw to the downhill sawyer by grasping one handle and the middle of the saw blade with the teeth facing away from sawyer.

Usually the uphill sawyer (the primary sawyer who will finish the cut) lubricates the saw and positions a guiding hand on the back of the saw for the first few strokes.

If you are going to roll the severed log out of the way, be sure to make a compound cut. The goal is for the sections of log to have as little surface resistance against each other as possible. The larger the log, the more careful the planning needed for the compound cut. Make the cuts where you will be safe and you will be able to move the log.

Your dominant hand (power hand) should be placed firmly around the saw's handle. Your other hand can rest on top of the handle to guide the saw and to help maintain your balance. Your dominant hand pulls the saw straight back to the side of your body. Sawyers often grip the saw too tightly with their guiding hand. This tends to pull the saw across their body.

ALWAYS pull—NEVER push!

Allow your partner to pull. Pushing may cause the saw to buckle.

As one sawyer pulls, the other sawyer keeps a relaxed grip on the handle. The sawyer neither pushes nor holds back. Holding back is called riding the saw, which makes it harder for the other sawyer to pull.

If you momentarily relax your grip, the saw will reposition itself in your hand for the pull stroke. Relaxing your grip also increases circulation in your hands and reduces fatigue.

If one sawyer needs to change body position (to drop to a kneeling position, for instance), the other sawyer needs to adjust the angle of the saw to accommodate the change.

Wedges should be placed as soon as there is room behind the back of the saw to insert them. For long logs, two wedges usually are inserted at the 10 and 2 o'clock positions and driven firmly until they are snug. If the wedges are not snug, the saw could be damaged.

Be sure the saw travels into and out of the kerf in a straight line. Look down the saw toward the other sawyer.

If the log is going to be finished up by single bucking, whenever **EITHER** sawyer determines it is time to stop sawing, both sawyers **STOP**. Do not allow your judgment to be swayed by your partner even if it means more single bucking will need to be done. Leave the downhill side whenever you feel you are in jeopardy. Each partner MUST honor the request of the other without pressuring the other partner.

If the log is going to be severed by double bucking (on flat terrain) be sure that the circle of the saw remains parallel to the ground. Do not have one end higher than the other.

Usually the bottom bark has not been removed. Carefully look at the shavings. When they change to the color of the bark, the log has been severed and only the bark is holding it. If the log falls on mineral soil, the impact can force rocks into the bark. The rocks can dull the saw's teeth. Usually the cut is stopped when wood-colored fibers are no longer being removed.

When the cut is finished, or when it is being finished by single bucking, remove the handle on the downhill side of the saw and allow the uphill sawyer to pull the saw free. Make sure the downhill sawyer is in a safe location before the uphill sawyer continues the cut.

DO NOT remove the wedges before removing the saw. The wedges may be holding the log in position. When wedges are removed in these situations, the severed log shifts, binding the saw. If the wedges are loose enough to be lifted straight up, it is safe to do so—do not wiggle them out. Once the saw is free, the wedges can be safely removed from the uphill side. Be prepared for the log to move.

Felling

Safety Considerations

Safety considerations for felling apply whether you are using chain saws or crosscut saws. The tree and the forces acting on it cannot tell the difference between handtools and power tools.

Chapter 4—Crosscut Saw Tasks and Techniques

Situational Awareness—Analyze the felling job by considering:
- Species (live or dead).
- Size and length.
- Soundness or defects.
- Twin tops.
- Widow makers or hangups.
- Heavy branches or uneven weight distribution.
- Spike tops.
- Splits and frost cracks.
- Deformities, such as mistletoe.
- Damage by lightning or fire.
- Heavy snow loading.
- Bark soundness.
- Direction of lean.
- Degree of lean (slight or great).
- Type of lean (head or side lean).
- Nesting or feeding holes.
- Punky (swollen and sunken) knots.
- Rusty (discolored) knots.
- Frozen wood.
- Footing.

Analyze the base of the tree for:
- Thud (hollow) sound when struck.
- Conks and mushrooms.
- Rot and cankers.
- Shelf or "bracket" fungi.
- Wounds or scars.
- Split trunk.
- Insect activity.
- Feeding holes.
- Bark soundness.
- Resin flow on bark.
- Unstable root system or root protrusions.

Examine surrounding terrain for:
- Steepness.
- Irregularities in the ground.
- Draws and ridges.
- Rocks.
- Stumps.
- Loose logs.
- Debris that can fly back or kick up at the sawyers.

Examine the immediate work area for:
- People, roads, or vehicles.
- Powerlines.
- Widow makers.
- Hangups.
- Other trees that may be affected.
- Other trees that may have to be felled first.
- Reserve (leave) trees.
- Structures.
- Openings to fall trees.
- Snags.
- Fire-weakened trees.
- Hazards such as trees, rocks, brush, or low-hanging limbs.

Walk out and thoroughly check the intended lay where the tree is supposed to fall. Look for dead treetops, snags, and widow makers that may cause kickbacks, allow the tree to roll, or cause another tree or limb to become a hazard.

The escape route and alternate routes must be predetermined paths where the sawyer can escape once the tree is committed to fall or has been bucked. Safety zones should be no less than 20 feet from the stump, preferably behind another tree that is sound and large enough to provide protection. Escape routes and safety zones should be 90 to 135 degrees from the direction of fall. Sawyers must select and prepare the work area and clear escape routes and alternate routes before starting the first cut.

Felling Sizeup

Most accidents are caused by falling debris. Watch overhead throughout the cut, glancing regularly at the saw, the kerf, and the top of the tree.

When you approach the tree to be felled, observe the top. Check for all overhead hazards that may come down during felling.

Look at the limbs. Are they heavy enough on one side to affect the desired felling direction? Do the limbs have heavy accumulations of ice and snow?

Are the limbs entangled with the limbs of other trees? If so, they can snap off or prevent the tree from falling after it has been cut.

Is the wind strong enough to affect the tree's fall? Wind speeds higher than 15 miles per hour may be strong enough to affect the tree's fall. If so, stop felling. Strong winds may blow over other trees and snags in the area. Erratic winds require special safety considerations.

Check all snags in the immediate area for soundness. A snag may fall at any time with a gust of wind, the vibration of a tree fall, or as the snag's roots succumb to rot. If it is safe to do so, begin by felling any snag in the cutting area that is a hazard.

Clear small trees, brush, and debris from the base of the tree. Remove all material that could cause you to trip or lose balance. Also remove material that will interfere with your use of the saw, wedges, and ax. Be careful not to fatigue yourself with unnecessary swamping. Remove only what is needed to work safely around the base of the tree and to provide escape routes.

The importance of sound holding (hinge) wood cannot be overemphasized. Determine the condition of the holding wood by sounding it with an ax. Look up while doing so, in case any debris is dislodged. Check for frost cracks or other weak areas in the holding wood. The desired felling direction can be adjusted to compensate for weak areas in the holding wood. The depth of the undercut can also be adjusted to take advantage of the holding area.

Most trees have two natural leans: the predominant head lean and the secondary side lean. The leaning weight of the tree will be a combination of these two leans. Both must be considered when determining the desired felling direction. The desired felling direction can usually be chosen within 45 degrees of the combined lean, provided there is enough sound holding (hinge) wood to work with, especially in the corners of the undercut.

Use a plumb bob or ax to evaluate the tree's lean. Project a vertical line up from the center of the tree's butt and determine if the tree's top lies to the right or left of the projected line.

A pistol-grip tree (see glossary) may appear to be leaning in one direction while most of the weight is actually in another.

Look at the treetop from at least two different spots at right angles to each other. This will be done again during the sizeup, but take every opportunity to determine the correct lean.

63

In summary, during felling sizeup:

- Observe the top.
- Check for snags.
- Swamp out the base.
- Assess the soundness of the holding wood.
- Assess the lean.

Establishing Escape Routes

Determine the escape routes. With the desired felling direction in mind, determine your escape route. Consider which side of the tree you will be making your final cut on and select a path that will take you at least 20 feet behind the stump when the tree begins to fall. Don't choose a path directly behind the tree. It is best to prepare two escape routes in case you switch your location on the final cut (figure 4-22).

Look for a large solid tree or rock for protection. The tree or rock must be at least 20 feet away from the stump and not directly behind it. Make sure that debris that could trip you is cleared from the escape route. Practice the escape.

Walk out the intended lay of the tree (figure 4-23). Look for any obstacles that could cause the tree to kick back over the stump or cause the butt to jump or pivot as the tree hits the ground. Look for any small trees or snags that could be thrown into your escape route. Check to be sure the cutting area is clear of people.

Reexamine the escape route. Using the observations you made walking out the lay, reexamine the escape route. Be sure that your chosen route will be the safest escape—before you begin to cut.

Figure 4-23—Check the intended lay of the tree for unwanted obstacles.

Placing the Undercut

After the escape routes are established, specific methods shall be used to cut the tree. We are only going to discuss the conventional undercut because of its broad application for all timber types and because it provides a solid foundation from which to learn additional cutting techniques.

Before beginning the undercut, prepare the tree for cutting (figure 4-24). Thick bark should be removed to:
- Keep the saw sharp.
- Make wedges more effective.
- See how the cuts are lining up.

It takes three cuts to fell a tree. Two cuts form the undercut (or face cut) and the third forms the backcut. The correct relationship of these cuts results in safe and effective tree felling. Before discussing the felling procedure, we will analyze the mechanics of the felling cuts. Undercutting and backcutting construct the hinge that controls the direction and fall of the tree.

The undercut serves three purposes. First, it allows the tree to fall in a given direction by removing the tree's support in the direction of the face. Second, it enables control because the tree slips off, rather than jumps off, the stump. Third, when the tree is breaking the holding wood, the tree is prevented from kicking back.

The undercut determines where the tree will fall. The undercut can be made by:

- Chopping out the entire undercut with an ax.
- Making the undercut with a crosscut saw.
- Making the horizontal cut with a crosscut saw and chopping the face out with an ax.

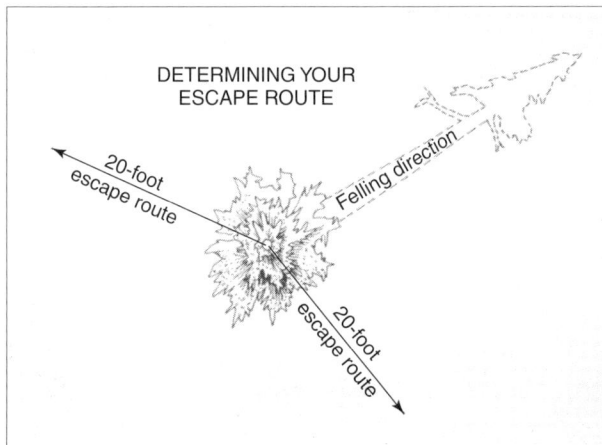

Figure 4-22—Keep the felling direction in mind when planning escape routes.

Chapter 4—Crosscut Saw Tasks and Techniques

Using a Crosscut Saw and an Ax—Making the horizontal cut with a saw and chopping out the undercut with an ax uses these tools to their best advantage. The horizontal saw cut is put in first, allowing the sawyer to more easily place a level cut. The ax helps keep the cut free of any dutchmans.

Making the Undercut With a Saw—This method is not recommended for several reasons:

• A high degree of skill is required to have both cuts meet exactly. When the cuts don't meet exactly, they create a dutchman. Careful ax work **MUST** be used to clean out the dutchman.

• Saws do not function well when they are used to cut diagonally.

Observe overhead hazards and look up often during the undercut (figure 4-25).

The tree is faced in the general direction of the tree's lean. Ideally, the undercut is made in the same direction as the tree's lean. Depending on structures, roads, other trees, trails, and compliance with predetermined leads, the desired felling direction may be to one side or the other of the lean. Normally, a desired direction is chosen less than 45 degrees from the lean.

Figure 4-24—An ax can be used to remove the bark.

Chopping Out the Undercut With an Ax—Although this approach may appear to be the hardest, it has advantages in certain situations. If this method is used, the cut should be level so the backcut, which is parallel to it, will also be level.

• Chopping out the undercut is about as fast as sawing smaller trees.

• Chopping out the undercut may be best in restricted areas where one side of the tree does not offer standing room for the sawyer or does not have adequate clearance for the end of the saw.

• Chopping out the undercut will allow the sawyer to limit the number of cuts. It is hard to apply saw oil to the bottom edge of a saw in the horizontal felling position. This is especially useful if the tree is extremely pitchy.

• Chopping out the undercut is a good alternative when the saw handles cannot be vertical, or when a stiff bucking saw is used, or for any combination of factors that lead to an uncomfortable sawing position.

Figure 4-25—Hazard trees need to be removed to prevent anyone from working under them.

Chapter 4—Crosscut Saw Tasks and Techniques

The horizontal cut is a level cut. This cut is made close to the ground unless a snag is being felled or another factor creates special hazards for the sawyer. The horizontal cut dictates the direction of fall if the relationships of the three cuts are maintained. If there is any danger from above, such as snags, the cutting should be done while standing so the sawyer can watch the top and escape more quickly. After selecting the desired felling direction, estimate one-third the tree's diameter and begin the horizontal cut.

The specific direction of the undercut is determined by "gunning" the saw. Place the back of the saw against the back of the undercut. The direction saw's teeth point in the direction the tree should fall. Short snags sometimes require an undercut deeper than one-third the tree's diameter to offset the tree's balance. Trees with heavy leans may not allow you to insert the horizontal cut as deep as one-third of the tree's diameter without pinching the saw.

When the horizontal cut is complete, remove the bark from an area on both sides of the kerf. The bark can be removed with your ax. Watch out in case the ax glances off the bole.

The sloping cut needs to be angled so that when the face closes the tree is fully committed to your planned direction of fall. As the face closes, the holding wood breaks. If the holding wood breaks and the tree is still standing straight, the tree could fall away from the desired direction.

A general rule for the sloping cut is a 45-degree angle. Remember that it is important that the face not close until the tree is fully committed to your planned direction of fall (figure 4-26).

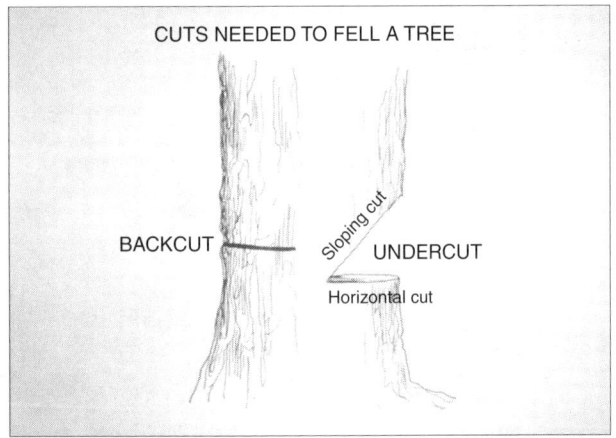

Figure 4-26—An undercut and a backcut are required to fell a tree. A horizontal cut and a sloping cut make up the undercut. The backcut is the third cut needed to fell a tree.

Lining up the sloping cut with the horizontal cut so that they meet, but do not cross, is one of the most difficult tasks in felling. When the cuts cross, a dutchman is formed (figure 4-27). If the

Figure 4-27—When the sloping cut and the horizontal cut cross, a dutchman is formed.

tree were felled with a dutchman, first the dutchman would close, then the tree would split vertically (barber chair), or the holding wood would break off. Felling control would be lost. A weak tree might snap off somewhere along the bole or at the top. It is difficult to make the sloping cut and the horizontal cut meet correctly on the opposite side of the tree. This is because the sawyer cannot look behind the tree while sawing.

Practicing on high stumps will help you become skilled at lining up these cuts.

The holding wood is the wood immediately behind the undercut. The most important portion of the holding wood is in the very corners of the cut (the first 4 to 8 inches inside the bark). The horizontal and sloping cut must not overlap in this region. If they do, the undercut must be cleaned up so no dutchman is left in these corners. Care must be taken not to cut the undercut too deeply while cleaning up. This will affect the amount of room available for wedges.

If the sloping cut is so shallow that cleaning it up will create too deep of an undercut, stop the sloping cut directly above the end of the horizontal cut.

The undercut needs to be cleaned out. Any remaining wood will cause the face to close prematurely and the holding wood will be broken behind the closure.

Once the face has been cleaned, recheck the felling direction. Place the saw back in the face and check the gunning or stick an ax head into the face and look down the handle. The back of the undercut should be perpendicular to the desired felling direction.

If the tree is not aimed in the direction that you want it to fall, extend the horizontal and sloping cuts as needed, maintaining a single plane for each of the two cuts.

Each sawyer pulls alternately. Sawyers **NEVER PUSH.** Pushing the saw can cause it to buckle. The saw is pulled directly back to the side of the sawyer with a slight upward arc at the end of the stroke.

The sawyer who is not pulling relaxes the grip on the handle and allows the hands and arms to be moved at the saw's speed to the position where the next pull stroke begins.

Cutting the Backcut

The third cut needed to fall a tree is the backcut. The relationship of this cut to the face is important for proper tree positioning and the sawyer's safety. The backcut can be made from either side of the tree. Choose the safest side to cut on, (not under any lean, good escape route, and so forth).

The best way to envision these cuts is by the use of a rectangle. The bottom corner is the back of the horizontal cut. The opposite upper corner will be the back of the backcut (figure 4-28).

The height of the rectangle is referred to as the stump shot. It is an antikickback device to prevent the tree from kicking back over the stump if it hits another tree on its fall. This is especially important to sawyers who are felling trees through standing timber.

The width of the rectangle is the holding wood. As the backcut is made, the sawyer must be sure not to cut this wood. Maintaining the holding wood is the key to safe and effective felling.

Hold the saw level so that the backcut will be level when the cut is complete. You want to be sure that when the cut is finished it will line up with the top corner of the opposite rectangle. If the cut is angled, wedging power and the height of the stump shot could be altered.

Keep at least three wedges and an ax readily accessible while making the backcut. Keep the ax within arm's reach. The size of wedge depends on tree diameter. For a 24-inch tree, a good combination would be two 10- to 12-inch wedges and one 4- to 6-inch wedge.

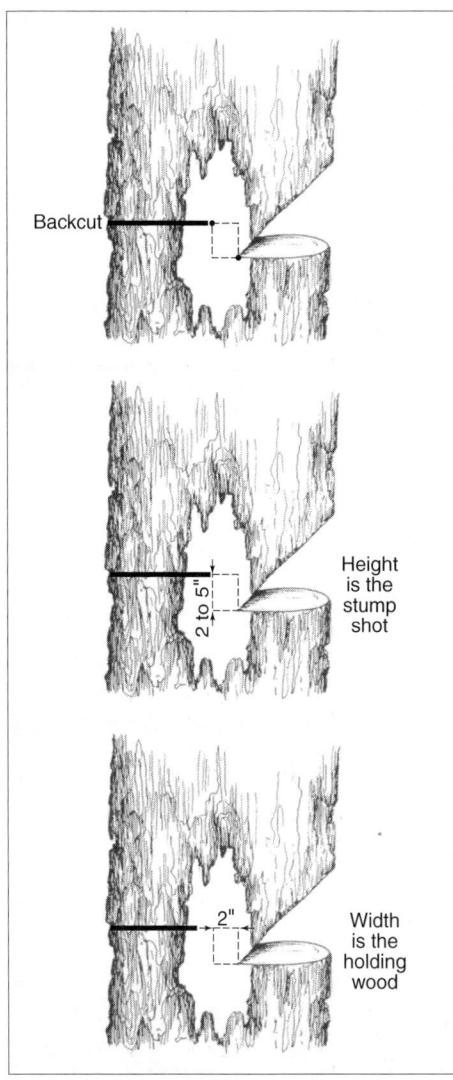

Figure 4-28—An imaginary rectangle can help the sawyer understand the importance of the backcut.

If there is any wind at all, two wedges are recommended. The second wedge lends stability. With only one wedge, the tree can set up a rocking action between the holding wood and that wedge. A strong wind could tear out the holding wood.

Remove thick bark immediately above and below the backcut's kerf where wedges will be placed. The bark could compress, lessening the lifting power of the wedges. The wedges should be spread to better stabilize the tree in case of erratic winds.

Chapter 4—Crosscut Saw Tasks and Techniques

If two sawyers are working together, the head sawyer will place the saw behind the tree where it will not block the escape route. Never take the saw with you along the escape route—it could slow you down.

When the second sawyer is watching from a safe location, the head sawyer can drive the wedges, causing the tree to lift and commit to fall. When the tree begins to move, the head sawyer can escape along the escape route from the stump.

Do not hesitate at the stump waiting for the tree to lift enough to clear a stuck saw. **LEAVE THE SAW.**

While in a safe location, both sawyers need to continue looking up for overhead hazards. There is a tendency to look at the tree as it hits the ground, leaving the sawyers unaware of limbs that may be thrown back from other trees near the stump. **LOOK UP.** If rocks or other material are dislodged when the tree hits the ground, yell a warning.